DATE DUE

FE 25 '98			
MR 18 '98			
MY 28 '98			

DEMCO 38-296

A Course in Group Theory

JOHN F. HUMPHREYS

Department of Pure Mathematics
University of Liverpool

Oxford New York Tokyo

OXFORD UNIVERSITY PRESS

1996

Riverside Community College
Library
4800 Magnolia Avenue
Riverside, California 92506
SEP '97

QA 177 .H85 1996

Humphreys, J. F.

A course in group theory

Walton Street, Oxford OX2 6DP
rd New York
nd Bangkok Bombay
Calcutta Cape Town Dar es Salaam Delhi
Florence Hong Kong Istanbul Karachi
Kuala Lumpur Madras Madrid Melbourne
Mexico City Nairobi Paris Singapore
Taipei Tokyo Toronto
and associated companies in
Berlin Ibadan

Oxford is a trade mark of Oxford University Press

Published in the United States by
Oxford University Press Inc., New York

© John Humphreys, 1996

All rights reserved. No part of this publication may be reproduced, stored in a retrieval system, or transmitted, in any form or by any means, without the prior permission in writing of Oxford University Press. Within the UK, exceptions are allowed in respect of any fair dealing for the purpose of research or private study, or criticism or review, as permitted under the Copyright, Designs and Patents Act, 1988, or in the case of reprographic reproduction in accordance with the terms of licences issued by the Copyright Licensing Agency. Enquiries concerning reproduction outside those terms and in other countries should be sent to the Rights Department, Oxford University Press, at the address above.

This book is sold subject to the condition that it shall not, by way of trade or otherwise, be lent, re-sold, hired out, or otherwise circulated without the publisher's prior consent in any form of binding or cover other than that in which it is published and without a similar condition including this condition being imposed on the subsequent purchaser.

A catalogue record for this book is available from the British Library

Library of Congress Cataloging in Publication Data

ISBN 0 19 853453 1 (Hbk)
ISBN 0 19 853459 0 (Pbk)

Typeset by the author using LaTeX
Printed in Great Britain by
Biddles Ltd, Guildford and King's Lynn

For my parents

Preface

This book provides a first course in the theory of finite groups. The main objective is to discuss various classification problems. To start with, there are the most obvious classifications, such as how many groups of a given order are there? The theory is developed to enable answers to be given to this question in increasingly more complicated situations. This type of investigation culminates in the complete list of groups with 31, or fewer, elements in Chapter 22. However, on the way to this result, it will be seen that classification problems of different types are more appropriate to group theory. The key role played by the arithmetic nature of the order of the group soon becomes apparent. Thus, if p and q are primes, questions such as: determine the groups with p^2 elements, or determine groups with pq elements, appear as suitable problems for analysis.

This process takes on a further dimension once the Jordan–Hölder Theorem is established. This shows the way in which a finite group G is constructed from a sequence of 'simple' groups (known as the 'composition factors' of G). These simple groups may be viewed as the atoms of finite group theory. In one of the most remarkable achievements of human endeavour, the complete list of the finite simple groups was obtained in the early 1980s. One of the main objectives of this book is to be able to outline this result, and so give the reader some idea of the complexity and depth of this classification.

The exposition is meant to be almost self-contained, with just a small number of prerequisities from number theory and matrix theory. These are collected together in Appendix A. The pace of the exposition is deliberately slow, with most details of proofs written out in full. There are also many illustrative examples given to elucidate definitions and proofs. There is a summary given at the end of each chapter to draw together the main ideas discussed. These summaries are intended to replace the more detailed chapter introductions given in many other textbooks. There seems to be more to be gained in trying to place results in context after the basic definitions have been absorbed than before. Many of the summaries contain remarks on the history of the subject. For the most part, these have been drawn from the book by van der Waerden (1985).

These methods of presentation are used in recognition of the fact that modern undergraduates find abstract algebra more difficult than their predecessors. Changes in the pre-university mathematics syllabuses and methods of learning mean that very much less algebra is taught than once was the case. The pace of exposition and level of exercises are intended to build confidence (and enjoyment) in this exciting and rewarding area of mathematics.

This book grew out of an undergraduate course taught in the University of Liverpool over several years. The first 18 chapters have all, at one time or another, been included in this course. There is plenty of material here on which to base a semester introduction to finite groups. The remaining chapters have all been taught in first year postgraduate courses.

The book is divided into 25 chapters. The first of these is entirely introductory, and simply gives definitions and examples. Chapter 2 is concerned with the general theory of maps between sets with the objective of defining the symmetric group $S(X)$. This chapter also introduces the idea of a relation on a set in order to consider the additive group of congruence classes modulo n. In Chapter 3, elementary consequences of the definitions are discussed.

The next sequence of four chapters is concerned with the concept of subgroup. This is introduced in Chapter 4, where the basic idea of the group generated by a subset is also discussed. Chapter 5 introduces cosets in order to obtain a proof of Lagrange's Theorem and some of its elementary consquences. At first sight, Chapter 6 may seem a diversion from this progression since it is an elementary introduction to the theory of error-correcting codes. These codes were introduced in order to ensure that information may be accurately transmitted. However, it turns out that the decoding process for the linear codes is, in fact, nothing more than a left coset decomposition. In Chapter 7, the idea of normal subgroups is introduced, and the way in which these may be used to construct quotient groups is described.

Chapter 8 is concerned with homomorphisms of groups. These are maps between groups which preserve the group multiplication. The basic result here is the Homomorphism Theorem. Its two main consequences are the First and Second Isomorphism Theorems. In Chapter 9, we return to the idea of permutations first considered in Chapter 2, and present the basic results in this area.

The next three chapters are intended to introduce the Sylow results. Chapter 10 discusses G-sets; these are sets X which admit an action by elements of a group G, the basic result here being the Orbit–Stabiliser Theorem, which says that for each $x \in X$ the number of distinct elements of X obtained by acting on x by elements of G (the orbit of x) is equal to the quotient of the number of elements in G by the number of elements of G which fix x (the stabiliser of x). This chapter ends with a classification of groups of order p^2 (where p is a prime). The Sylow theory is discussed in Chapter 11. For a finite group G of order $p^n k$, where p does not divide k, these results establish the existence of subgroups of G with p^n elements.

Chapter 12 gives several examples of applications of the Sylow results to classification problems.

The next objective is to investigate finite abelian groups. This is the subject of Chapter 14. Before this can be done, it is necessary to discuss direct products in some detail, and this is achieved in Chapter 13. This chapter discusses internal direct products as well as two important general constructions, the central product and the pullback.

The main objective in Chapter 15 is the Jordan–Hölder Theorem. This result shows the importance of the composition factors in determining the structure of a finite group. It is shown in Chapter 16 that the composition factors are, in fact, simple groups. It is in this sense that we understand the finite simple groups as the building blocks from which other finite groups are constructed.

The next pair of chapters are concerned with the notion of soluble groups. These were introduced by Galois in connection with his investigations of the solubility of polynomial equations. The concept is defined in Chapter 17 and basic properties of soluble groups are obtained. In Chapter 18, it is shown that a finite p-group is soluble and also that all finite groups with fewer than 60 elements are soluble.

The next four chapters develop the theory of extensions. This is to enable one to address the problem of determining the isomorphism classes of groups G with a given normal subgroup N whose quotient group G/N is isomorphic to a given group H. Chapter 19 considers the special case of semidirect products when we may suppose that G has a subgroup isomorphic to H. The basic results on extension theory are then given in Chapter 20. In Chapter 21, we discuss two important special types of extension: central extensions (when N is in the centre of G) and cyclic extensions (when H is cyclic). Finally in this sequence, the list of all isomorphism types of groups with 31 or fewer elements is obtained. These groups are listed in Appendix B.

The final three chapters of the book discuss some of the finite simple groups. In Chapter 23, the simple groups $PSL(n, q)$ associated with $n \times n$ matrix groups over a finite field with q elements are introduced. Chapter 24 contains an introduction to the Mathieu groups, which are examples of the sporadic simple groups. Finally, in Chapter 25 an introduction to the classification of finite simple groups is given, without any proofs.

As well as the Appendices already mentioned, complete solutions to the exercises are provided. This is part of the basic philosophy of the book. Group theory requires an active reader. Exercises are included in order to test and develop the basic concepts of the subject. Reading the text without writing anything or trying any problems is unlikely to lead to real understanding. The problems are designed with students in mind, and most of the earlier exercises have been set to Liverpool undergraduates. These exercises are intended to build confidence and skill.

I wish to thank John Brinkman and Francis Coghlan for reading several drafts of the text with care and insight. I am grateful for their comments. I also wish to thank my wife Patricia, not just for the support and encouragement she gave during the writing of the book, but also for her accurate and helpful proof-reading. Finally I should like to thank Elizabeth Johnston and Julia Tompson of Oxford University Press for their patient help and advice.

Liverpool J. F. H.

January 1996

Contents

1

Definitions and examples

In this chapter, we shall give a few basic definitions and then concentrate on examples of groups. This will be done for two reasons: to familiarise the reader with the basic ideas of the subject, and to provide a catalogue from which we can illustrate definitions and theorems in subsequent chapters.

Definition 1.1 *A* group *is a set G together with an operation $\circ$ satisfying the following requirements:*

(G1) *for each pair x, y of elements of G, $x \circ y$ is an element of G* (closure axiom);

(G2) *for all elements x, y, z of G, $(x \circ y) \circ z = x \circ (y \circ z)$* (associativity axiom);

(G3) *there is an element e in G such that for all g in G*

$$e \circ g = g = g \circ e \qquad \text{(identity axiom);}$$

(G4) *given an element g in G, there is an element $g^\star$ in G such that*

$$g \circ g^\star = e = g^\star \circ g \qquad \text{(inverse axiom).}$$

Definition 1.2 *A group G is said to be* infinite *if the number of elements in the set G is infinite, otherwise the group is* finite.

Definition 1.3 *A group G is* abelian *if for all x and y in G, $x \circ y = y \circ x$.*

We now start to consider our list of examples.

Example 1.4 Let G be the set $\mathbf{Z}$ of integers and the operation $\circ$ be addition of numbers. Then $\mathbf{Z}$ satisfies the four axioms under the operation $+$ if we take the element e in (G3) to be the integer 0, and the inverse $x^\star$ in (G4) to be the negative $-x$. In fact this is an infinite abelian group. Other familiar number systems, such as the set $\mathbf{Q}$ of rational numbers, the set $\mathbf{R}$

of real numbers and the set **C** of complex numbers are also infinite abelian groups with the operation of addition.

Example 1.5 Considering the operation of multiplication on the set **Z**, we see that the first three axioms are satisfied, provided we take e to be the number 1. However, the fourth axiom fails in two ways. In the first place, zero has no inverse: there is no integer $0^\star$ such that $0 \circ 0^\star = 1$. Secondly, although there is a number $2^\star$ with $2 \circ 2^\star = 1$ (take $2^\star$ to be $1/2$), there is no *integer* satisfying this condition. If we consider each of the number systems **Q**, **R** and **C** under multiplication, zero has no inverse, but each other number does. In fact the sets $\mathbf{Q}^\times, \mathbf{R}^\times$ and $\mathbf{C}^\times$ of non-zero elements of **Q**, **R** and **C**, respectively, are infinite abelian groups under the operation of multiplication.

Example 1.6 We can provide examples of finite abelian groups as follows. Let n be an integer greater than 1 and let ω be the complex number $e^{2\pi i/n}$ so that $\omega^n = 1$. (Here i denotes one of the complex square roots of -1, and e denotes the base of natural logarithms.) Let G be the set of distinct complex numbers $1, \omega, \ldots, \omega^{n-1}$. Define an operation on G by the rule $\omega^j \omega^k = \omega^{j+k}$, so that the operation is the same as multiplication of complex numbers. Checking the closure axiom here is not completely obvious, since we must check that ω^{j+k} is an element of the group, so that we can write ω^{j+k} as one of the complex numbers $1, \omega, \ldots, \omega^{n-1}$. Since $0 \leq j, k \leq n-1$, there are only two possibilities: either $j + k < n$, in which case ω^{j+k} is in G, or $j + k$ is greater than $n - 1$, in which case $j + k$ is of the form $n + t$, for some $t < n$. Since $\omega^n = 1$, it then follows that

$$\omega^j \omega^k = \omega^{j+k} = \omega^{n+t} = \omega^n \omega^t = \omega^t.$$

Taking the identity to be 1 and the inverse of ω^j to be ω^{n-j}, we see that G is a finite group with n elements. Since

$$\omega^j \omega^k = \omega^{j+k} = \omega^{k+j} = \omega^k \omega^j,$$

it may be seen that this group is abelian. In fact, this is an example of a *cyclic* group, this being the name given to a group in which each element is a power of one particular element of the group. We shall therefore denote this group by C_n. This notation will also be used later for any cyclic group with n elements.

Example 1.7 Let $M_n(\mathbf{R})$ denote the set of $n \times n$ matrices with real entries. This is a group under addition of matrices if we take e to be the zero $n \times n$ matrix (all entries 0) and the inverse of the matrix X to be $-X$. The set is not a group under matrix multiplication. This is because the identity for matrix multiplication is the identity matrix

$$I_n = \begin{pmatrix} 1 & 0 & . & . & . & 0 \\ 0 & 1 & . & . & . & 0 \\ . & & & . & & . \\ . & & & & . & . \\ . & & & & . & . \\ 0 & 0 & . & . & . & 1 \end{pmatrix}.$$

The multiplicative inverse of a matrix A is then the matrix A^{-1} in the usual sense of matrix theory. The condition that A has an inverse is that A has non-zero determinant, so not all matrices are invertible. The subset $GL(n, \mathbf{R})$ of invertible elements in $M_n(\mathbf{R})$ is a group under multiplication, taking e to be I_n and the inverse of X to be the matrix inverse X^{-1}. Replacing $\mathbf{R}$ by other number systems such as $\mathbf{Q}$ or $\mathbf{C}$, we obtain other groups of matrices under multiplication, namely $GL(n, \mathbf{Q})$ and $GL(n, \mathbf{C})$.

Example 1.8 Let G consist of the following four 2×2 matrices:

$$I = \begin{pmatrix} 1 & 0 \\ 0 & 1 \end{pmatrix}, \; A = \begin{pmatrix} 1 & 0 \\ 0 & -1 \end{pmatrix}, \; B = \begin{pmatrix} -1 & 0 \\ 0 & 1 \end{pmatrix}, \; C = \begin{pmatrix} -1 & 0 \\ 0 & -1 \end{pmatrix}.$$

We claim that these form a group under matrix multiplication. The easiest way to justify this claim is to calculate the table of all possible products.

	I	A	B	C
I	I	A	B	C
A	A	I	C	B
B	B	C	I	A
C	C	B	A	I

The table checks closure because each entry is in the set G. We can also see that I is the identity element and that each of I, A, B and C is its own inverse. Thus the table checks each of the group axioms apart from the associativity axiom, (G2). However, matrix multiplication is well known to be associative, and so this axiom is automatically satisfied.

Example 1.9 We next consider the groups of symmetries of a geometric object in a plane. The elements of this group are transformations of the plane which leave the geometric figure invariant. This abstract example will be illustrated by the case when the geometric figure is an equilateral triangle. To simplify the subsequent discussion, we will label the vertices of the triangle with the integers 1, 2, 3, as shown.

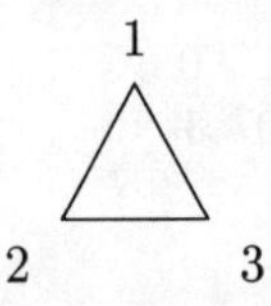

There are several transformations of the plane which preserve the triangle. For example, rotating the plane 120 degrees anticlockwise through the centre of gravity of the triangle will do this. We can then rotate through a further 120 degrees. Another rotation through this angle will return the triangle to its original position. This is considered to be the same as no rotation, and so is regarded as the identity element. Three further symmetries are obtained by reflecting the triangle in each of the perpendicular bisectors of the sides. Thus the reflection in the bisector of the side joining vertex 2 to vertex 3 leaves 1 fixed but interchanges the vertices 2 and 3. In fact these six symmetries are the only possible symmetries of the triangle. We label them as follows:

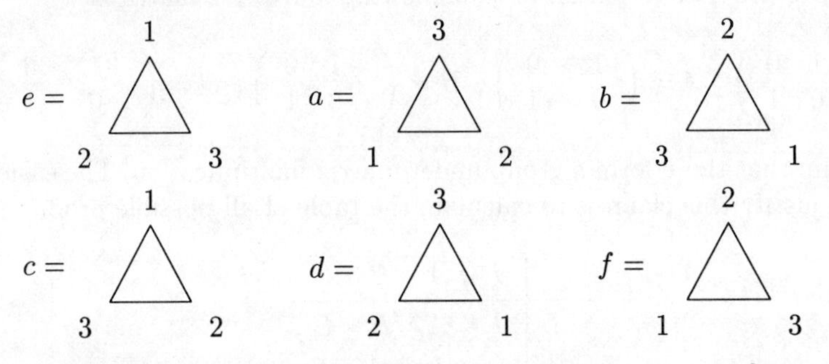

We define an operation on these symmetries by saying that the symmetry $x \circ y$ should be obtained by first performing y and then applying x to the result. Since a denotes rotation through 120 degrees, and c denotes reflection in the perpendicular bisector of the 'base' line, $a \circ c$ is reflection in the bisector of the side joining vertex 1 to vertex 2 ($= f$). In this way, it can be checked that the table for the set under this operation is as follows:

	e	a	b	c	d	f
e	e	a	b	c	d	f
a	a	b	e	f	c	d
b	b	e	a	d	f	c
c	c	d	f	e	a	b
d	d	f	c	b	e	a
f	f	c	d	a	b	e

This table checks each of the group axioms except the associativity axiom (G2). However, composition of transformations is associative, as will be shown in Chapter 2, so this is a group.

In a similar way, for any odd integer n with $n \geq 3$, we can construct the group of symmetries of a regular n-sided polygon. It turns out that this group has $2n$ elements (n rotations and n reflections in the perpendicular bisectors of the sides). When n is even, the group of symmetries of a regular n-sided polygon also has $2n$ elements, but these are n rotations and n reflections (either in the diagonals or in the line joining the midpoints of opposite sides). This group is known as the *dihedral group* and is denoted here by $D(n)$, and also referred to as D_n, or even D_{2n}, by other authors. As we shall see later, the properties of this group often depend on the parity of n. It is also possible to work out symmetry groups of other plane figures and also of three-dimensional objects.

In this notation, the group of symmetries of an equilateral triangle is the dihedral group $D(3)$ and is a non-abelian group with six elements.

Example 1.10 In this example, we give a general way to construct new groups from given groups G and H. The *direct product of G and H* is the set of ordered pairs

$$G \times H = \{(g, h) : g \in G, h \in H\},$$

under the operation

$$(g_1, h_1) \circ (g_2, h_2) = (g_1 \circ_G g_2, h_1 \circ_H h_2),$$

where $\circ_G$ and $\circ_H$ denote the operations in G and H, respectively. It may easily be checked that $G \times H$ is a group under this operation. For example, if $G = H = C_2 = \{1, -1\}$, $G \times H$ has four elements $(1,1), (1,-1), (-1,1)$ and $(-1,-1)$ and multiplication table

	$(1,1)$	$(1,-1)$	$(-1,1)$	$(-1,-1)$
$(1,1)$	$(1,1)$	$(1,-1)$	$(-1,1)$	$(-1,-1)$
$(1,-1)$	$(1,-1)$	$(1,1)$	$(-1,-1)$	$(-1,1)$
$(-1,1)$	$(-1,1)$	$(-1,-1)$	$(1,1)$	$(1,-1)$
$(-1,-1)$	$(-1,-1)$	$(-1,1)$	$(1,-1)$	$(1,1)$

Notation To conclude this chapter, we remark that for the most part, from now on, we will normally use *multiplicative notation* for groups. In this notation, we omit the composition sign $\circ$, and write the operation as though it were multiplication (so replace $x \circ y$ by xy). We also write the identity of G as 1 (in the cases where confusion with the number 1 might occur, we may write the identity of G as 1_G) and write the inverse of g as g^{-1}. Thus the group axioms, written in this notation, appear as:

(G1) for each pair x, y of elements of G, xy is an element of G,

(G2) for all elements x, y, z of G, $(xy)z = x(yz)$,

(G3) there is an element 1 in G such that for all g in G

$$1g = g = g1,$$

(G4) given an element g in G, there is an element g^{-1} in G such that

$$gg^{-1} = 1 = g^{-1}g.$$

When dealing with abelian groups, it will sometimes be convenient to use *additive notation* in which the group operation is written as $+$, the identity as 0 and the inverse of g as $-g$.

Summary for Chapter 1

In this chapter, we introduce some of our standard groups. These include:

the additive groups of numbers **Z, Q, R** and **C**;

the multiplicative groups $\mathbf{Q}^\times, \mathbf{R}^\times$ and $\mathbf{C}^\times$ of non-zero elements of **Q**, **R** and **C**, respectively;

the finite multiplicative group C_n of powers of the complex number $\omega = e^{2\pi i/n}$;

the additive group $M_n(\mathbf{R})$ of $n \times n$ matrices with real entries;

the multiplicative group $GL(n, \mathbf{R})$ of $n \times n$ real matrices with non-zero determinant;

the group $D(n)$ of symmetries of a regular n-sided polygon.

We also discuss the direct product of groups G and H. This is the set of ordered pairs

$$G \times H = \{(g, h) : g \in G, h \in H\}$$

under the operation

$$(g_1, h_1)(g_2, h_2) = (g_1 g_2, h_1 h_2).$$

The first person to use the word 'group' in a technical sense was the French mathematician Évariste Galois (1811–1832). In his short life, Galois intoduced this branch of mathematics as a result of his investigations into solutions of polynomial equations by radicals. We shall have more to say about Galois's contributions to the origins of the subject later. His death was caused by a wound received in a duel. Cauchy also introduced group theory at about the same time as Galois (but independently of him), as a result of the studies Cauchy was making of the symmetries of functions. It happened that Galois's terminology was adopted, since it was much shorter than Cauchy's.

Exercises 1

1. Determine which of the following sets are groups under the specified operations:

(a) the integers under the operation of subtraction;

(b) the set $\mathbf{R}$ of real numbers under the operation $\circ$ given by

$$a \circ b = a + b + 2;$$

(c) the set of odd integers under the operation of multiplication;

(d) the set of $n \times n$ real matrices whose determinant is either 1 or -1, under matrix multiplication.

2. Calculate the multiplication table for the following eight 2×2 complex matrices, and deduce that they form a non-abelian group:

$$I = \begin{pmatrix} 1 & 0 \\ 0 & 1 \end{pmatrix},\ A = \begin{pmatrix} i & 0 \\ 0 & -i \end{pmatrix},\ B = \begin{pmatrix} -1 & 0 \\ 0 & -1 \end{pmatrix},\ C = \begin{pmatrix} -i & 0 \\ 0 & i \end{pmatrix},$$

$$D = \begin{pmatrix} 0 & 1 \\ 1 & 0 \end{pmatrix},\ E = \begin{pmatrix} 0 & i \\ -i & 0 \end{pmatrix},\ F = \begin{pmatrix} 0 & -1 \\ -1 & 0 \end{pmatrix},\ G = \begin{pmatrix} 0 & -i \\ i & 0 \end{pmatrix}.$$

3. Find the multiplication table for the eight symmetries of a square.

4. Find the symmetry groups of

(a) a non-square rectangle,

(b) a parallelogram with unequal sides which is not a rectangle,

(c) a non-square rhombus.

5. Write down the multiplication tables for the groups $C_2 \times C_3$ and $C_3 \times C_3$.

6. Show that $G \times H$ is abelian if and only if G and H are each abelian.

2

Maps and relations on sets

In this chapter, we consider two situations which lead to further examples of groups. The first of these arises from the general theory of maps between sets. The reader who feels confident with these ideas should proceed directly to Definition 2.17.

Definition 2.1 *Let X and Y be sets. A* map *(or* function*) $f : X \to Y$ is a rule which assigns to each element x of X an element $f(x)$ of Y.*

Definition 2.2 *A map $f : X \to Y$ is* injective *(or* one-to-one*) if f takes distinct values on distinct elements of X:*

$$\text{if } x_1 \neq x_2 \text{ then } f(x_1) \neq f(x_2).$$

(Equivalently, f is injective if the fact that $f(x_1) = f(x_2)$ implies that $x_1 = x_2$.)

Definition 2.3 *A map $f : X \to Y$ is* surjective *(or* onto*) if each y in Y appears as a value of f:*

$$\text{for all } y \in Y, \text{ there exists an element } x \in X \text{ such that } y = f(x).$$

Definition 2.4 *The map $f : X \to Y$ is* bijective *(or a* one-to-one correspondence*) if f is both injective and surjective.*

Remark It follows from the definition that the condition for two maps $f : X \to Y$ and $g : X \to Y$ to be equal is that $f(x) = g(x)$ for all x in X. It also follows that two maps $f : X \to Y$ and $g : X \to Z$ can only be equal if $Y = Z$.

Example 2.5 Among the most familiar maps are the functions studied in calculus. These are often defined by formulae. For example, when X and Y are both the set $\mathbf{R}$, examples of maps include:

$$f_1(x) = \sin x;$$

$$\begin{aligned} f_2(x) &= e^x; \\ f_3(x) &= x^3 - x; \\ f_4(x) &= 2x + 1. \end{aligned}$$

The map f_1 is neither injective (for example $f_1(0) = f_1(\pi)$) nor surjective (there is no x with $f_1(x) > 1$); the map f_2 is injective but not surjective (since e^x is never negative); f_3 is surjective but not injective (since $f_3(0) = f_3(1)$); and f_4 is bijective.

Such functions are often represented pictorially, particularly when X and Y are both $\mathbf{R}$, by drawing their graphs. We can interpret our ideas using the graph of the function as follows: to say that f is injective means that each horizontal line $y = c$ intersects the graph in at most one place; and to say that the function is surjective means that each horizontal line $y = c$ intersects the graph in at least one place.

Example 2.6 We shall be mainly interested in the case when the sets X and Y each have a finite number of elements. In this case it is less usual to define a map by a formula, and more customary to define a map by giving a list of its values. For example, suppose that X is a set with two elements $\{a, b\}$ and that $Y = \{u, v\}$. Then for any map f from X to Y we need to know $f(a)$ and $f(b)$. There are therefore four such maps specified by

$$f_1(a) = u,\ f_1(b) = v; \qquad f_2(a) = u,\ f_2(b) = u;$$

$$f_3(a) = v,\ f_3(b) = v; \quad \text{and} \quad f_4(a) = v,\ f_4(b) = u.$$

Of these, the maps f_2 and f_3 are clearly neither injective nor surjective, whereas f_1 and f_4 are both bijective.

Definition 2.7 *Given maps $f : X \to Y$ and $g : Y \to Z$ the* composite map $g \circ f : X \to Z$ *is defined by $g \circ f(x) = g(f(x))$ for all $x \in X$.*

Example 2.8 Consider the maps $f, g : \mathbf{R} \to \mathbf{R}$ defined by $f(x) = \sin x$ and $g(x) = 2x + 1$. Then $f \circ g$ is the map defined by $f \circ g(x) = \sin(2x + 1)$ whereas $g \circ f(x) = 2 \sin x + 1$.

Example 2.9 When $X = Y = \{a, b\}$ the four maps from X to X are

$$f_1(a) = a,\ f_1(b) = b; \quad f_2(a) = a,\ f_2(b) = a;$$

$$f_3(a) = b,\ f_3(b) = b; \quad \text{and} \quad f_4(a) = b,\ f_4(b) = a.$$

The composition of these maps is as shown in the following table:

	f_1	f_2	f_3	f_4
f_1	f_1	f_2	f_3	f_4
f_2	f_2	f_2	f_2	f_2
f_3	f_3	f_3	f_3	f_3
f_4	f_4	f_3	f_2	f_1

Note that this table shows that f_1 is an identity element for this set of maps. However, the set is not a group since neither of the elements f_2 nor f_3 has an inverse. We shall discuss the question of inverses of maps later.

One of the most often quoted facts about maps is the fact that their composition is an associative operation.

Proposition 2.10 *Given maps $f : X \to Y$, $g : Y \to Z$ and $h : Z \to W$ the composite maps $(h \circ g) \circ f : X \to W$ and $h \circ (g \circ f) : X \to W$ are equal.*

Proof By definition, for all $x \in X$,

$$\begin{aligned} ((h \circ g) \circ f)(x) &= (h \circ g)(f(x)) \\ &= h(g(f(x))) \\ &= h(g \circ f(x)) \\ &= (h \circ (g \circ f))(x). \end{aligned}$$

It follows that the maps $(h \circ g) \circ f$ and $h \circ (g \circ f)$ are equal. □

We next discuss the idea of the inverse of a map.

Definition 2.11 *For any set X, the* identity map *is the map $id_X : X \to X$ defined by $id_X(x) = x$ for all $x \in X$. It is clear from the definition that if $f : X \to Y$ is any map, then $f \circ id_X = f$ and $id_Y \circ f = f$.*

Definition 2.12 *Given a map $f : X \to Y$, we say that f has an* inverse map *if there is a map $g : Y \to X$ such that $g \circ f = id_X$ and $f \circ g = id_Y$.*

Example 2.13 In Example 2.9, we saw that the maps f_1 and f_4 are each their own inverses, but that neither of f_2 and f_3 have inverses.

Proposition 2.14 *If a map $f : X \to Y$ has an inverse, then this inverse is unique. It will be denoted by f^{-1}.*

Proof Suppose that g and h are both inverses for f. Thus

$$g \circ f = h \circ f = id_X \quad \text{and} \quad f \circ g = f \circ h = id_Y.$$

Then

$$\begin{aligned} h = h \circ id_Y &= h \circ (f \circ g) \\ &= (h \circ f) \circ g \quad \text{by Proposition 2.10} \\ &= id_X \circ g = g, \end{aligned}$$

as required. □

Proposition 2.15 *A map $f : X \to Y$ has an inverse if and only if f is a bijection.*

Proof Suppose first that f is a bijection. Define a map $g : Y \to X$ by the rule:

$$g(y) = x \quad \text{if and only if} \quad f(x) = y.$$

Note that g is a map since there is at most one element of X assigned to any y by g (since f is injective), but every element of Y is assigned to some x in X (since f is surjective). It follows directly from the definition that the composite maps $f \circ g$ and $g \circ f$ are the appropriate identity maps.

Conversely, suppose that f has an inverse function f^{-1}. We first show that f is injective. Suppose that $f(x_1) = f(x_2)$. Apply the map f^{-1} to both sides to obtain

$$x_1 = id_X(x_1) = f^{-1}(f(x_1)) = f^{-1}(f(x_2)) = id_X(x_2) = x_2,$$

so that f is injective. To show that f is surjective, take any y in Y. Then

$$y = id_Y(y) = f \circ f^{-1}(y) = f(f^{-1}(y)) = f(x)$$

where $x = f^{-1}(y)$. This completes the proof. □

Corollary 2.16 *If $f : X \to Y$ and $g : Y \to Z$ are bijections, then so is the map $g \circ f$.*

Proof Using Proposition 2.15, f and g have inverse maps f^{-1} and g^{-1}. To establish the result, it is sufficient to observe that

$$(g \circ f)(f^{-1} \circ g^{-1}) = id_Z \quad \text{and} \quad (f^{-1} \circ g^{-1})(g \circ f) = id_X,$$

so that the inverse of $g \circ f$ is $f^{-1} \circ g^{-1}$. Then Proposition 2.15 shows that $g \circ f$ is a bijection. □

Definition 2.17 *A bijection* π *from a set* X *to itself is a* permutation on X.

Corollary 2.18 *Let* X *be any set. Then the set* $S(X)$ *of permutations* $f : X \to X$ *is a group under composition of functions.*

Proof We check the group axioms for $S(X)$. Closure follows by Corollary 2.16. The associativity axiom is then established by Proposition 2.10. The identity axiom follows if we take the identity element to be the map id_X. Finally, the definition of an inverse function shows that if f^{-1} is the inverse of f then f is the inverse of f^{-1}, and so f^{-1} is a bijection. □

Notation From now on, as for general groups, we shall omit the composition sign $\circ$ and write the composite of permutations $\pi \circ \rho$ simply as juxtaposition $\pi\rho$, referring to composition as multiplication.

The case when the set X consists of the three integers $\{1, 2, 3\}$ will now be considered in detail.

Example 2.19 First note that there are 27 possible maps from the set $X = \{1, 2, 3\}$ to itself. Most of these are not injective, but six of these maps are permutations. Denote by $S(3)$ the set of all possible permutations of the set of three integers 1, 2 and 3. The six permutations may be denoted as follows:

$$\epsilon = \begin{pmatrix} 1 & 2 & 3 \\ 1 & 2 & 3 \end{pmatrix}, \ \alpha = \begin{pmatrix} 1 & 2 & 3 \\ 2 & 3 & 1 \end{pmatrix}, \ \beta = \begin{pmatrix} 1 & 2 & 3 \\ 3 & 1 & 2 \end{pmatrix},$$

$$\gamma = \begin{pmatrix} 1 & 2 & 3 \\ 1 & 3 & 2 \end{pmatrix}, \ \delta = \begin{pmatrix} 1 & 2 & 3 \\ 3 & 2 & 1 \end{pmatrix}, \ \phi = \begin{pmatrix} 1 & 2 & 3 \\ 2 & 1 & 3 \end{pmatrix},$$

where the values of each map are given in the second row. Thus the map β takes 1 to 3, $\beta(2) = 1$ and $\beta(3) = 2$. Since the maps are injective, the entries in the second row are all distinct. To calculate $\beta\gamma$, for example, work out what this map does to each of 1, 2 and 3. Thus

$$(\beta\gamma)(1) = \beta(\gamma(1)) = \beta(1) = 3;$$
$$(\beta\gamma)(2) = \beta(\gamma(2)) = \beta(3) = 2;$$
$$(\beta\gamma)(3) = \beta(\gamma(3)) = \beta(2) = 1.$$

Hence $\beta\gamma = \delta$. In this way, we can check that the six elements of $S(3)$ multiply together according to the following table:

	ϵ	α	β	γ	δ	ϕ
ϵ	ϵ	α	β	γ	δ	ϕ
α	α	β	ϵ	ϕ	γ	δ
β	β	ϵ	α	δ	ϕ	γ
γ	γ	δ	ϕ	ϵ	α	β
δ	δ	ϕ	γ	β	ϵ	α
ϕ	ϕ	γ	δ	α	β	ϵ

The reader should check the above entries before going any further. This shows that $S(3)$ is a non-abelian group with six elements.

This example has a strong relationship with Example 1.9. Although the objects in these two groups are quite different, the tables are the 'same' in the sense that if we replace each entry in this table by its corresponding roman letter, we obtain the table of Example 1.9. We say that the groups are *isomorphic* (literally 'have the same shape') in this case. More formally, two groups G and H are said to be isomorphic if there is a bijection ϕ from G to H such that corresponding entries in the two tables are equal:

$$\text{for all } x \text{ and } y \text{ in } G, \quad \phi(x \circ y) = \phi(x) \circ \phi(y).$$

We now consider the other main idea of this chapter.

Definition 2.20 *A* relation *on a set X is a subset R of the cartesian product*

$$X \times X = \{(x_1, x_2) : x_1, x_2 \in X\}.$$

We shall often write xRy to mean that (x, y) is in R.

Example 2.21 Some standard examples are the relations $<, >$ $\leq$ and $\geq$ on number systems such as $\mathbf{R}$. So, for example, if R is the relation $<$, then $(3, 5)$ is in R because $3 < 5$, but $(4, 3)$ is not in R because $4 \not< 3$.

Example 2.22 Given a map $f : X \to X$, we can define a relation R to be the subset of elements of the form $(x, f(x))$. This is also known as the *graph* of f.

Example 2.23 Let n be an integer greater than 1. The relation on $X = \mathbf{Z}$ of *congruence modulo* n is defined by xRy if and only if

$$x \equiv y \mod n.$$

Thus, two integers x and y are congruent modulo n if and only if x and y have the same remainder when divided by n. When $n = 2$ there are two congruence classes of integers, those with no remainder when divided by 2 (the even integers) and those with remainder 1 when divided by 2 (the odd integers).

Definition 2.24 *A relation R on X is an* equivalence relation *on X if R satisfies the following three requirements: for all x, y, z in X,*

(1) *xRx (reflexive property);*
(2) *if xRy then yRx (symmetric property); and*
(3) *if xRy and yRz then xRz (transitive property).*

Example 2.25 None of the relations $<, >, \leq$ and $\geq$ on $\mathbf{R}$ is an equivalence relation because none of them is symmetric.

Example 2.26 The relation of congruence modulo n is an equivalence relation. To show this, we check the three requirements:

(1) since n divides 0 ($0 = 0 \times n$), $x \equiv x \bmod n$;
(2) if $x \equiv y \bmod n$, then n divides $x - y$ so that n divides $y - x$ and hence $y \equiv x \bmod n$;
(3) if $x \equiv y \bmod n$ and $y \equiv z \bmod n$, then n divides $x - y$ and also $y - z$ and so n divides $x - y + y - z = x - z$, so $x \equiv z \bmod n$.

Definition 2.27 *Given an equivalence relation R on a set X, the* equivalence class *$[x]_R$ of an element x in X is the set of elements of X related to x:*

$$[x]_R = \{y \in X : (x, y) \in R\}.$$

Remark Because an equivalence relation is symmetric, it does not matter whether we write $(x, y) \in R$ or $(y, x) \in R$ in Definition 2.27.

Definition 2.28 *For any set X, a set of non-empty subsets of X is a* partition *of X if each element of X is in precisely one of the subsets. It follows that the union of the set of subsets is X, but the intersection of any two different subsets is the empty set.*

Proposition 2.29 *The equivalence classes of any equivalence relation on a set X form a partition of the set.*

Proof First note that because R is reflexive, any element x in X is in the equivalence class $[x]_R$ and so the union of all equivalence classes is X. If z is in both $[x]_R$ and $[y]_R$, then xRz and yRz. Since R is symmetric, we see that zRy and so the fact that R is transitive means that xRy. Thus y is in $[x]_R$, and so xRy and yRx, so that $[x]_R = [y]_R$. □

Example 2.30 In the case when we consider the relation of congruence modulo n, we usually refer to *congruence classes* rather than equivalence classes. The congruence class of x will be denoted $[x]_n$, although the subscript, and even the square brackets, may be omitted when the meaning is unambiguous. When $n = 2$, there are two congruence classes: $[0]_2$ consisting of all the even integers and $[1]_2$ consisting of all the odd integers. It is possible to define addition and multiplication of congruence classes by

$$[x]_n + [y]_n = [x + y]_n \text{ and } [x]_n[y]_n = [xy]_n.$$

These rules are deceptively simple, and need to be considered carefully. The notation $[x]_n$ is a shorthand notation for infinitely many integers (all those with the same remainder as x when divided by n). We need, therefore, to understand why the addition and multiplication are 'well-defined'. This will be a recurring problem whenever we define an operation on a set of equivalence classes by giving its effect in terms of representatives of those equivalence classes. In the present context, we can formulate the well-defined property for addition as:

$$\text{if } [x_1]_n = [x_2]_n \text{ and } [y_1]_n = [y_2]_n, \text{ then } [x_1 + y_1]_n = [x_2 + y_2]_n,$$

and the property for multiplication as

$$\text{if } [x_1]_n = [x_2]_n \text{ and } [y_1]_n = [y_2]_n, \text{ then } [x_1y_1]_n = [x_2y_2]_n.$$

These facts are established in the following result.

Proposition 2.31 *Addition and multiplication on the set* $\mathbf{Z}_n$ *of congruence classes modulo n are well-defined.*

Proof Suppose that $[x_1]_n = [x_2]_n$ and $[y_1]_n = [y_2]_n$, so that n divides $x_1 - x_2$ and also n divides $y_1 - y_2$. Thus

$$x_1 - x_2 = nr \text{ and } y_1 - y_2 = ns \text{ for some integers } r \text{ and } s.$$

Then

$$(x_1 + y_1) - (x_2 + y_2) = (x_1 - x_2) + (y_1 - y_2) = n(r + s)$$

so that n divides $(x_1 + y_1) - (x_2 + y_2)$ and so

$$[x_1 + y_1]_n = [x_2 + y_2]_n,$$

as required.

The proof for multiplication is left to the reader. □

Corollary 2.32 *The set $\mathbf{Z}_n$ is an abelian group under the operation of addition of congruence classes.*

Proof The group axioms are easily checked, with the identity element being $[0]_n$ and the inverse of $[x]_n$ being $[-x]_n$. □

Example 2.33 In $\mathbf{Z}_3$, $[0]_3$ denotes the set of integers divisible by 3, $[1]_3$ denotes those integers, such as 10 and -5, which have remainder 1 when divided by 3, and $[2]_3$ denotes those integers with remainder 2 when divided by 3. Thus $\mathbf{Z}_3$ has the following addition table:

	$[0]_3$	$[1]_3$	$[2]_3$
$[0]_3$	$[0]_3$	$[1]_3$	$[2]_3$
$[1]_3$	$[1]_3$	$[2]_3$	$[0]_3$
$[2]_3$	$[2]_3$	$[0]_3$	$[1]_3$

Summary for Chapter 2

In this chapter two general situations are considered. The first of these concerns properties of maps between sets. The key ideas here are that a map $f : X \to Y$ is *injective* if no two distinct elements of X are mapped by f to the same element of Y; f is *surjective* if every element of Y is the image under f of some element of X, and f is *bijective* if f is both injective and surjective. It is shown that composition of maps is an associative operation, and also that a necessary and sufficient condition for a map to have an inverse is that the map is a bijection. A bijective map f from a set X to itself is called a *permutation*, and it is shown that the set of permutations of a set X is a group under composition of maps. The case when $X = \{1, 2, 3\}$ is studied in some detail.

The other general situation concerns relations on sets. The basic idea here is that a relation R on a set X is an *equivalence relation on* X if R satisfies the following three requirements:

(1) for all x in X, xRx (reflexive property);
(2) if xRy then yRx (symmetric property); and
(3) if xRy and yRz then xRz (transitive property).

The standard example of an equivalence relation is the relation of congruence modulo an integer n. For any positive integer n, the set $\mathbf{Z}_n$ of congruence classes modulo n is a group under addition of congruence classes.

If R is an equivalence class on a set X, it is also shown that the set of equivalence classes of R partitions the set X.

From the historical point of view, groups of permutations were the first types of groups to be studied. The axiomatic definition given in 1.1 was only formulated over many years, with contributions from many mathematicians including Cayley, Dyck and Weber.

Exercises 2

1. Let $X = \{a, b, c\}$ and $Y = \{u, v\}$. List all the maps from X to Y and list all the maps from Y to X.

2. Let $g : X \to Y$ and $f : Y \to Z$ be functions. Show that

(a) if f and g are both injective then fg is injective;
(b) if f and g are both surjective then fg is surjective.

Give examples to show that if f is injective and g is surjective then fg need neither be injective nor surjective.

3. When $X = \{a, b, c\}$, list all the maps $f : X \to X$ which are constant (so that $f(a) = f(b) = f(c)$). Write down the composition table for these maps. Do these maps form a group?

4. Prove that the relation on the set $\mathbf{Z}$ defined by xRy if $x + y$ is an even integer is an equivalence relation, and determine the equivalence classes. Is the relation xRy if $x + y$ is divisible by 3 an equivalence relation?

5. Write down the addition table for the congruence classes modulo 4, and the multiplication table for the non-zero congruence classes modulo 5.

6. Show that multiplication of congruence classes modulo n is well-defined.

3

Elementary consequences of the definitions

In this chapter, we shall consider some logical consequences of the group axioms. For example, if we are given a group G, the third axiom ensures the existence of an identity element in G. There seems to be nothing in the statement of the axioms to prevent a group having more than one identity element. However, as we shall now show, this cannot be the case.

Proposition 3.1 *In any group G there is only one identity element.*

Proof Suppose G had two identity elements, 1 and e, say. What we mean by this is that, for all g in G,

$$1g = g = g1 \quad \text{and} \quad eg = g = ge.$$

Apply the first of these equations with $g = e$ to obtain that $1e = e$. Take $g = 1$ in the second set of equations to obtain $1 = 1e$. Putting these together gives $1 = 1e = e$, so that $e = 1$. □

In a similar way, the fourth group axiom does not say that a given group element could not possibly have more than one inverse. We now consider the possible elements which could be inverses for a given element g of a group G.

Proposition 3.2 *Every element in a group has a unique inverse.*

Proof Suppose that g is an element of a group G with two inverses $g^\star$ and g^{-1}, say. This means that

$$gg^\star = 1 = g^\star g \tag{1}$$

and

$$gg^{-1} = 1 = g^{-1}g. \tag{2}$$

Then

$$
\begin{aligned}
g^\star(gg^{-1}) &= g^\star 1 \quad \text{using (2)} \\
&= g^\star \quad \text{using axiom (G3).}
\end{aligned}
$$

Also

$$
\begin{aligned}
g^\star(gg^{-1}) &= (g^\star g)g^{-1} \quad \text{using axiom (G2)} \\
&= 1g^{-1} \quad \text{using (1)} \\
&= g^{-1} \quad \text{by (G3).}
\end{aligned}
$$

We have therefore shown that $g^\star = g^{-1}$, so that there is a unique inverse for the arbitrary group element g. □

Remark We have seen this result for inverses previously in the special case when the elements g are permutations (Proposition 2.14).

Propositions 3.1 and 3.2 are special cases of a general result which enables equations to be solved in groups.

Proposition 3.3 *Let a and b be elements of a group G. There is a unique element x in G such that $ax = b$, and there is also a unique element y in G such that $ya = b$.*

Proof We shall establish the first of these claims, leaving the other to the reader. There are two steps in the proof, one to show that there is an element x satisfying $ax = b$, the other to show that there is only one such element. Taking uniqueness first, suppose the equation has two solutions x and z, say, so that

$$ax = b = az.$$

By axiom (G4), a has an inverse a^{-1}. Multiply both sides of the equation $ax = az$ on the left by a^{-1} to get

$$a^{-1}(ax) = a^{-1}(az).$$

Now apply associativity (G2) to obtain

$$(a^{-1}a)x = (a^{-1}a)z.$$

Axiom (G4) gives that $a^{-1}a = 1$, so we deduce that $1x = 1z$. Applying axiom (G3) then gives that $x = z$, so that the equation has a unique solution.

We now show that the equation has at least one solution by checking that the element $x = a^{-1}b$ is an element of G and that this value of x satisfies the equation $ax = b$. Since, by hypothesis, a is in G, a^{-1} is in G by (G4) and so $a^{-1}b$ is also in G by axiom (G1). Also

$$\begin{aligned} ax = a(a^{-1}b) &= (aa^{-1})b \quad \text{by (G2)} \\ &= 1b \quad \text{by (G4)} \\ &= b \quad \text{by (G3).} \end{aligned}$$

This completes the proof. □

Remark 1 Care must be taken in the definition of x in the final paragraph of the above proof. The temptation to interchange orders of products must be resisted. For example, $x = a^{-1}b$ is NOT a solution of $xa = b$, so that the solutions x and y of Proposition 3.3 need not be equal. The best way to ensure correct proofs is to be able to justify each and every step by a group axiom or a rule of reasoning, as we have done in the above proof. Of course, once confidence has been built up it will not be necessary to write all these justifications down so that excessive details can then be omitted.

As an illustration of this process, let a, b be elements of a group G. Consider the problem of finding the solution x of the equation $bxa^{-1} = a^{-1}b$ in G. To do this, first multiply the equation on the left by b^{-1} to obtain $xa^{-1} = b^{-1}a^{-1}b$. Multiply this equation on the right by a to find that $x = b^{-1}a^{-1}ba$. Note that this value for x does not 'simplify' in any way.

Remark 2 Proposition 3.3 has an interpretation in the multiplication table for G. Since ax_1 and ax_2 are in the same row of the table, the uniqueness part of the proposition shows that no group element occurs more than once in any row of the table. The fact that a solution of $ax = b$ exists for each b shows that each group element does occur in each row. Hence each group element occurs once and only once in each row (and by a similar argument in each column).

Remark 3 Taking $b = a$ in Proposition 3.3 proves Proposition 3.1 (that the identity element is unique). Proposition 3.2 is also a special case of Proposition 3.3 obtained by taking $b = 1$.

The next step is to give two further consequences of Proposition 3.3.

Corollary 3.4 *Let a and b be elements of a group G. Then the inverse of ab is $b^{-1}a^{-1}$. (Note reversal of order.)*

Proof Consider the product

$$\begin{aligned}(ab)(b^{-1}a^{-1}) &= a(b(b^{-1}a^{-1})) \text{ by (G2)} \\ &= a((bb^{-1})a^{-1}) \text{ by (G2) again} \\ &= a(1a^{-1}) \text{ by (G4)} \\ &= aa^{-1} \text{ by (G3)} \\ &= 1 \text{ by (G4).}\end{aligned}$$

For any g in G, the equation $gx = 1$ has a unique solution. We know g^{-1} is a solution of this equation, so the above argument shows that the inverse of ab is indeed $b^{-1}a^{-1}$. □

Corollary 3.5 *Let g be an element of a group G. The inverse of g^{-1} is g.*

Proof The equations

$$gg^{-1} = 1 = g^{-1}g$$

say that g^{-1} is the inverse of g, and also that g is the inverse of g^{-1}. □

We now investigate a further consequence of the associativity axiom (G2). Let a, b, c and d be elements of a group G. We can form a variety of products such as $a(b(cd)), (ab)(cd), (a(bc))d$ and $((ab)c)d$. Repeated use of (G2) shows that these are, in fact, all equal. We would usually make use of this 'fourfold' associativity law without comment and write all the possible ways to bracket the terms as $abcd$, except when a particular use of brackets reveals useful relations between the group elements. There are similar n fold associative rules for any n, which we will now prove. The reader may wish to omit the proof of this fact on first reading.

Proposition 3.6 *Let $g_1, g_2, \ldots$ be elements of a group G and n be a positive integer. Define the set $\mathcal{P}_n(g_1, g_2, \ldots, g_n)$ inductively by*

$$\mathcal{P}_1(g_1) = \{g_1\}; \qquad \mathcal{P}_2(g_1, g_2) = \{g_1g_2\};$$

$$\mathcal{P}_n(g_1, g_2, \ldots, g_n) = \{xy : x \in \mathcal{P}_r(g_1, \ldots, g_r) \text{ and } y \in \mathcal{P}_s(g_{r+1}, \ldots, g_{r+s}) \text{ with } n = r + s\}.$$

Then $\mathcal{P}_n(g_1, g_2, \ldots, g_n)$ consists of a unique element of G which we denote $g_1g_2 \cdots g_n$. (Note that $\mathcal{P}_n(g_1, g_2, \ldots, g_n)$ consists of all the products formed *by bracketing the elements $g_1, g_2, \ldots, g_n$ in any way, keeping the factors in that order. The proposition says that these products are all equal as elements of G.)*

Proof The proof is by induction on n, the cases when $n = 1$ and $n = 2$ being clear. Let $g = xy$ be in $\mathcal{P}_n$ with $x \in \mathcal{P}_r$ and $y \in \mathcal{P}_s$. If $r > 1$, we write x as $g_1 z$ where $z = g_2 \ldots g_r$, by induction. Then, using induction again,

$$xy = (g_1 z)y = g_1(zy) = g_1(g_2 \ldots g_n).$$

If $r = 1$, induction gives that

$$xy = g_1 y = g_1(g_2 \ldots g_n).$$

Thus in either case, xy is equal to the single element $g_1(g_2 \ldots g_n)$. It follows that $\mathcal{P}_n(g_1, g_2, \ldots, g_n)$ consists of a single element. □

We next use induction to provide a definition of the powers of group elements.

Definition 3.7 *Let g be an element of a group G. Define g^0 to be 1, g^1 to be g, g^2 to be gg and, inductively, when n is positive, g^n to be gg^{n-1}. We can also define g^n when n is negative to be the inverse of g^{-n}.*

In an arbitrary group G, the powers of the elements of G satisfy the familiar index laws. The proof of these laws will turn out to be more difficult than might be anticipated.

Proposition 3.8 *Let g be an element of a group G and r, s be integers. Then*

(1) $g^r g^s = g^{r+s} = g^s g^r$;
(2) $(g^r)^s = g^{rs}$; *and*
(3) $g^{-r} = (g^{-1})^r = (g^r)^{-1}$, *so that the inverse of g^r is g^{-r}.*

Proof We will give the proof of (1) in detail. Consider first the case when r and s are both positive and proceed by induction on r. If $r = 1$, $gg^s = g^{1+s}$ by definition. The inductive hypothesis is that $g^r g^s = g^{r+s}$. Now

$$g^{r+1} g^s = (gg^r)g^s = g(g^r g^s) = gg^{r+s} = g^{r+s+1},$$

as required.

Now consider the case when one of r or s is zero. If $r = 0$,

$$g^0 g^s = 1g^s = g^s = g^{0+s}.$$

The proof when $s = 0$ is similar.

We next suppose that r and s are both negative. The idea now is to use the definition of negative powers together with the case already established when the indices are both positive to complete the proof in this case. The formal argument is as follows. By definition,

$$\begin{aligned} g^r g^s &= (g^{-r})^{-1}(g^{-s})^{-1} \\ &= (g^{-s}g^{-r})^{-1} \quad \text{by Corollary 3.4} \\ &= (g^{-s-r})^{-1} \quad \text{since } -r \text{ and } -s \text{ are positive} \\ &= (g^{-r-s})^{-1} \quad \text{since } -r-s=-s-r \\ &= g^{r+s} \quad \text{by definition.} \end{aligned}$$

Since $r+s=s+r$, it follows from this proof that $g^r g^s = g^s g^r$.

We are therefore left with the cases when one of r and s is positive and the other is negative. The proofs in these two cases are very similar, so we consider the case when $r>0$ and $s<0$, leaving the other case to the reader. There are three possibilities to consider.

$r+s>0$. By the case when both indices are positive,

$$g^{r+s}g^{-s} = g^{(r+s)-s} = g^r.$$

Since g^{-s} is the inverse of g^s by definition, we can multiply both sides by g^s to obtain $g^{r+s} = g^r g^s$, as required.

$r+s=0$. In this case $s=-r$, so that g^s is, by definition, the inverse of g^r and so $g^r g^s = 1 = g^0$.

$r+s<0$. Using the case where the indices are both positive,

$$g^{-(r+s)}g^r = g^{-r-s+r} = g^{-s}.$$

Multiplying both sides by g^{-r}, the inverse of g^r, gives

$$g^{-(r+s)} = g^{-s}g^{-r}.$$

Now take inverses of both sides to obtain the result.

The strategy of the proof of (2) is similar to the above, and the details are left to the reader. For (3), suppose first that r is positive, so that $g^{-r}=(g^r)^{-1}$ by definition. An easy inductive proof shows that $(g^{-1})^r$ is the inverse of g^r, as required. If $r=0$, then

$$g^{-r} = (g^{-1})^r = (g^r)^{-1} = 1.$$

Finally, if r is negative, $-r$ is positive and so we can again proceed by induction: when $r = -1$,

$$g^{-r} = g^1 = (g^{-1})^{-1} = (g^{-1})^r = (g^r)^{-1}.$$

Now suppose that

$$g^{-r} = (g^{-1})^r = (g^r)^{-1},$$

so that

$$\begin{aligned} g^{-r+1} &= g^{-r}g = (g^{-1})^r g \\ &= (g^{-1})^r(g^{-1})^{-1} = (g^{-1})^{r-1} \quad \text{using (1).} \end{aligned}$$

Also

$$\begin{aligned} g^{-r+1} &= (g^r)^{-1}g \quad \text{by induction} \\ &= (g^r)^{-1}(g^{-1})^{-1} \\ &= (g^{-1}g^r)^{-1} \quad \text{using Corollary 3.4} \\ &= (g^{r-1})^{-1} \quad \text{using (1).} \end{aligned}$$

This completes the proof. □

Remark The definition of g^2 when applied to a product such as $(xy)^2$ gives $(xy)(xy)$. Once again temptations to interchange orders must be resisted. This element is NOT usually equal to x^2y^2.

We next consider the idea of the order of an element g of a group G.

Definition 3.9 *Let g be an element of a group G. The* order *of g is the smallest positive integer n such that $g^n = 1$. If there is no positive integer such that $g^n = 1$, we say that g has* infinite order.

Remark 1 In any finite group, each element necessarily has finite order. To see this, consider the positive powers $g, g^2, g^3, \ldots$ and so on of any element g of the finite group G. Since these are all elements of the group and the group has a finite number of elements, this list must contain repetitions. Suppose that $g^r = g^s$ for some r and s with $r > s$. Then, since the inverse of g^s is g^{-s}, $g^{r-s} = g^s g^{-s} = 1$ and so g has finite order. An example of a group with an element of infinite order is provided by the group $GL(2, \mathbf{R})$ and the matrix $A = \begin{pmatrix} 1 & 1 \\ 0 & 1 \end{pmatrix}$. Since $A^n = \begin{pmatrix} 1 & n \\ 0 & 1 \end{pmatrix}$, A does not have finite order.

Remark 2 If g is an element of order n, it follows that g^{n-1} is the inverse of g. More generally, we can describe when any two powers of an element are equal.

Proposition 3.10 *Let g be an element of order n in a group G. Then $g^r = g^s$ if and only if n divides $r - s$. In particular $g^k = 1$ if and only if n divides k.*

Proof Suppose that n divides $r - s$, so that $r - s = nt$ for some integer t. Then

$$\begin{aligned} g^r = g^{s+nt} &= g^s g^{nt} \quad \text{using Proposition 3.8(1)} \\ &= g^s (g^n)^t \quad \text{using Proposition 3.8(2)} \\ &= g^s (1)^t \quad \text{since } g^n = 1 \\ &= g^s \quad \text{since } 1^t = 1. \end{aligned}$$

Conversely, suppose that $g^r = g^s$. By Proposition 3.8(3), the inverse of g^s is g^{-s}, so

$$g^{r-s} = g^r g^{-s} = g^s g^{-s} = 1.$$

Now write the integer $r - s$ in the form $qn + t$ with $0 \leq t < n$. The justification for doing this is known as the *division algorithm for integers*: given any integers a and b with $b > 0$, there exist integers c and d such that $a = cb + d$ where $0 \leq d < b$. This is Theorem A.1 in Appendix A. We then obtain

$$1 = g^{r-s} = g^{nq+t} = (g^n)^q g^t = 1^q g^t = g^t.$$

The definition of n (the smallest positive integer such that $g^n = 1$) forces t to be 0, so that n divides $r - s$, as required. □

To conclude this chapter, we show how the results on solving equations in groups can be used to determine the complete lists of groups with three elements and groups with four elements.

Groups with three elements

Suppose that G consists of the elements $1, a$ and b. Since 1 is the identity, the table starts as

	1	a	b
1	1	a	b
a	a	*	
b	b		

Consider the entry denoted $*$: it must either be 1 or b since a already occurs in that row. If it were 1, then the final entry in the first row would have to be b, giving two bs in the last column. It follows that the entry $*$ must be b. Using the fact that the columns contain each element precisely once, the table then fills in as

	1	a	b
1	1	a	b
a	a	b	1
b	b	1	a

We have therefore shown that any group with 3 elements must be isomorphic to the one with this table. The cyclic group, C_3, is an example of such a group taking $e^{2\pi i/3}$ to be a and $e^{4\pi i/3}$ to be b.

Groups with four elements

Let $G = \{1, a, b, c\}$. We can enter the first row and first column of the table using the fact that G has an identity element. Consider the possibilities for a^2.

(1) Suppose that $a^2 = 1$.

In this case, looking at the second row of the table, we see that ab must be b or c. However, b already occurs in the third column so $ab = c$ and $ac = b$. Similar reasoning gives the column containing a. We have the partial table

	1	a	b	c
1	1	a	b	c
a	a	1	c	b
b	b	c		
c	c	b		

There are two possibilities for b^2: 1 or a. For either of these choices, the table can be completed in a unique way. To check that these two possible tables are group tables, we note that the table with $b^2 = 1$ is (apart from a change of notation) the same as that in Example 1.10 and is isomorphic to $C_2 \times C_2$, while that with $b^2 = a$ is isomorphic to C_4, taking b to be i, a to be -1 and c to be $-i$.

(2) What about the situation if $a^2 \neq 1$?

We may suppose that no element of G has order 2 (since we could rename that element as a and apply the argument of (1)). If G had an element g of order 3, then $1, g$ and g^2 would be three of the elements of G. The fourth element would have to be its own inverse since we already know the inverses of $1, g$ and g^2. This means that the fourth element would have order 2 contrary to assumption. We conclude that in case (2), the group has an element g of order four, so $G = \{1, g, g^2, g^3\}$. But then g^2 would have order 2, so this case has already been considered.

We therefore obtain just two non-isomorphic groups with four elements, namely

	1	a	b	c
1	1	a	b	c
a	a	1	c	b
b	b	c	1	a
c	c	b	a	1

(this group is $C_2 \times C_2$ and is also known as the Klein four-group or Vierergruppe), and a group isomorphic to C_4 whose elements may be named as $1, g, g^2$ and g^3:

	1	g	g^2	g^3
1	1	g	g^2	g^3
g	g	g^2	g^3	1
g^2	g^2	g^3	1	g
g^3	g^3	1	g	g^2

Remark Many of the proofs of the results in this section do not utilise all the group axioms. For example, the proof of Proposition 3.6 works for any set G with an associative closed operation. Such a set is known as a semigroup. Examples of semigroups are provided by the set of maps from a set X to itself under the operation of composition of maps as discussed in Chapter 2. The index laws (1) and (2) of Proposition 3.8 also hold in any semigroup provided that r and s are positive. (A semigroup need not have an identity or inverses.) We shall only be concerned with the theory of groups.

Summary of Chapter 3

In this chapter, elementary logical consequences of the axioms are considered. Among the most important facts here are the result which enables equations of the type $ax = b$ to be solved (Proposition 3.3) and the index laws for powers of group elements (Proposition 3.8). The other important idea for later use is that of the *order* of an element g in a group G: the smallest positive integer k such that $g^k = 1$ (if it exists). This then gives a criterion for two powers of g to be equal (Proposition 3.10). A result which will be used repeatedly is that if g has order k and $g^r = 1$ then k divides r. The rules for solving equations in groups enable the complete list of multiplication tables of groups with three and four elements to be determined.

Exercises 3

1. Let G be a group in which $g^2 = 1$ for all g in G. Prove that G is abelian.

2. Let a, b and c be elements of the group G. Find the solutions x of the equations
(a) $axa^{-1} = 1$, (b) $axa^{-1} = a$, (c) $axb = c$ and (d) $ba^{-1}xab^{-1} = ba$.

3. Let G be a group and c be a fixed element of G. Define a new operation $*$ on G by

$$x * y = xc^{-1}y,$$

for all x and y in G. Prove that G is a group under the operation $*$.

4. List the orders of all the elements of the group $D(3)$ of Example 1.9.

5. Give an example of a group G with elements x and y such that $(xy)^{-1}$ is not equal to $x^{-1}y^{-1}$.

6. Let G be a group in which $(xy)^2 = x^2y^2$ for all x and y in G. Prove that G is abelian.

7. Let x and g be elements of a group G. Prove, using mathematical induction, that for all positive integers k,

$$(x^{-1}gx)^k = x^{-1}g^kx.$$

Deduce that g and $x^{-1}gx$ have the same order.

8. Let ω denote the complex number $e^{2\pi i/6}$, so that $\omega^6 = 1$. Let

$$X = \begin{pmatrix} \omega & 0 \\ 0 & \omega^{-1} \end{pmatrix}.$$

Show that $X^6 = I$ and calculate X^{-1}. Find a 2×2 matrix Y such that

$$XY = YX^{-1} \text{ and } Y^2 = X^3.$$

Show that the set $G = \{X^i,\ YX^j : 1 \leq i, j \leq 6\}$ with 12 elements is a group under matrix multiplication, and find the order of each element of G.

4

Subgroups

In this chapter, we introduce the concept of a subgroup of a group. The idea of considering subsets of an axiomatic system which also satisfy the axioms will be familiar if the reader has already met vector spaces (when the subsets are called subspaces), rings (subrings) or fields (subfields). We shall soon see that introducing this concept leads to new insights into group structures.

Definition 4.1 *A* subgroup *of a group G is a non-empty subset H of G which is itself a group under the same operation as that of G. There are two easy examples of subgroups in any group G: the subgroup $\{1\}$ and the subgroup G itself. We write $H \leq G$ to indicate that H is a subgroup of G, and $H < G$ to indicate that H is a subgroup which is not the group G.*

It would appear from our definition that we need to check each of the four group axioms to decide whether or not a given subset is a subgroup (checking the existence of an identity element will ensure the set is non-empty). However, there will be no need to check associativity (G2) in H. This is because if x, y and z are in H, then, since H is a subset of G, the elements x, y and z are automatically in G and so, since G is a group, $(xy)z = x(yz)$. In fact, we have the following result.

Proposition 4.2 *The following conditions on a subset H of group G are equivalent:*

(1) *H is a subgroup of G;*

(2) *H satisfies the three requirements:*

 (a) *the identity element of G is in H;*

 (b) *if x and y are in H, so is xy;*

 (c) *if h is in H, so is h^{-1};*

(3) *H satisfies the conditions:*

 (i) *the identity element of G is in H;*

 (ii) *if a and b are in H, so is ab^{-1}.*

Proof We show the equivalence of these statements as follows:
(1) $\Rightarrow$ (2) If H is a subgroup, it is non-empty and satisfies requirements (b) and (c). Also H will have an identity element 1_H such that $1_H h = h = h 1_H$ for all h in H. Since any element h in H is also in G, $h 1_G = h$, and so the uniqueness of solutions of equations (Proposition 3.3) shows that $1_G = 1_H$.

(2) $\Rightarrow$ (3) Condition (i) of (3) and condition (a) of (2) both say that the identity element of G is in H, so we only need to explain why (ii) holds. Let a and b be in H. By (c), b^{-1} is in H, so by (b), ab^{-1} is in H, as required.

(3) $\Rightarrow$ (1) We suppose that H satisfies (i) and (ii) and then check the group axioms for H. We have already seen that the associative law holds in H, and (i) ensures that H is non-empty and has an identity element. Let h be in H, and apply (ii) with $a = 1$ and $b = h$ to see that h^{-1} is in H, thus checking that the inverse axiom holds. Finally the group closure property holds since if x and y are in H, then x and y^{-1} are in H (as we have already seen). We now apply (ii) with $a = x$ and $b = y^{-1}$ to obtain that $x(y^{-1})^{-1} = xy \in H$, as required. □

Remark 1 We shall normally check that a subset is a subgroup by checking (2) in Proposition 4.2.

Remark 2 The proof of Proposition 4.2 shows that every subgroup of a group G contains the identity element of G.

Remark 3 Conditions (a) of (2) and (i) of (3) are only included to ensure that H is non-empty. If we know our subset is non-empty, then these become redundant. For example, if H had at least one element, applying (ii) of (3) with $a = b$ gives that 1 is in H.

Remark 4 If G is a finite group, it follows by Remark 2 before Proposition 3.10 that for any $g \in G$, g^{-1} is also of the form g^k for some positive integer k. It then follows by Remark 3 above that if H is a subset of G such that (i) H is non-empty and (ii) for all $x, y \in H, xy \in H$, then H is a subgroup of G. Conversely, any subgroup H of G satisfies the conditions (i) and (ii).

Example 4.3 Returning to the Examples 1.4 and 1.5, we see that $\mathbf{Z}$ is a subgroup of $\mathbf{Q}$ and in turn $\mathbf{Q}$ is a subgroup of $\mathbf{R}$. It follows easily from the definition that a subgroup of a subgroup is a subgroup, so we obtain a chain of additive abelian groups

$$\mathbf{Z} \leq \mathbf{Q} \leq \mathbf{R} \leq \mathbf{C},$$

and a chain of multiplicative groups

$$\mathbf{Q}^{\times} \leq \mathbf{R}^{\times} \leq \mathbf{C}^{\times}.$$

Example 4.4 In the group $GL(n, \mathbf{R})$ of invertible $n \times n$ matrices with real entries, the subset $SL(n, \mathbf{R})$ of matrices with determinant 1 is a subgroup. This follows since the identity matrix has determinant 1; if X and Y have determinant 1, so does XY; and the inverse of a matrix with determinant 1 exists and also has determinant 1.

Example 4.5 In any group of symmetries of a geometric figure in the plane, the subset of symmetries which are rotations about a fixed point is a subgroup. In the case of the group of symmetries of an equilateral triangle, there are three rotations and three reflections, and so this subgroup of rotations has three elements.

Proposition 4.6 *Let I be any index set and H_i $(i \in I)$ be subgroups of a group G. Then*

$$\bigcap H_i = \{g \in G : g \in H_i \text{ for all } i \in I\}$$

is a subgroup of G.

Proof Let $K = \bigcap H_i$. To show that K is a subgroup, we shall check the conditions of Proposition 4.2(2).

(a) Since each H_i is a subgroup, 1 is in H_i $(i \in I)$ and so 1 is in K.

(b) Suppose that x and y are in K. Then, for each i, x and y are in H_i, and so xy is in H_i, since H_i is a subgroup. Thus $xy \in K$, as required.

(c) Suppose that h is in K. Since h is in H_i $(i \in I)$ and H_i is a subgroup, h^{-1} is in H_i. Thus h^{-1} is in K. □

We now come to one of the most important ways to produce subgroups of a group G.

Definition 4.7 *Let X be any subset of a group G. Then $\langle X \rangle$, the* subgroup generated by X, *is the intersection of all the subgroups of G containing the set X. The fact that $\langle X \rangle$ is a subgroup follows from Proposition 4.6. Thus $\langle X \rangle$ is the smallest subgroup of G containing the set X.*

Notation When X is a list of elements $\{x_1, x_2, \ldots, x_n\}$ of G, we misuse notation and write $\langle x_1, x_2, \ldots, x_n \rangle$ for $\langle \{x_1, x_2, \ldots, x_n\} \rangle$. In the case when X consists of a single element of G, x say, the subgroup $\langle x \rangle$ is said to be *cyclic*. In that case the subgroup consists of all possible powers (positive, negative and zero) of x. In general, the elements of $\langle X \rangle$ are products of powers of the elements of X, since $\langle X \rangle$ necessarily contains all such products, and furthermore the set of all such products is a subgroup of G.

Example 4.8 Let G be the multiplicative group of powers of the complex number $\omega = e^{2\pi i/n}$ as discussed in Example 1.6. This group is cyclic and has n elements. It is for these reasons that we denote this group by C_n. It is easily seen that any two cyclic groups of the same order are isomorphic.

Example 4.9 Let G be the group $S(3)$ discussed in Example 2.19, and X be the permutations

$$\alpha = \begin{pmatrix} 1 & 2 & 3 \\ 2 & 3 & 1 \end{pmatrix} \text{ and } \gamma = \begin{pmatrix} 1 & 2 & 3 \\ 1 & 3 & 2 \end{pmatrix}.$$

Since $\langle X \rangle$ is a subgroup, and is closed under multiplication, it therefore contains α, α^2 ($= \beta$ in the notation of Chapter 2), $\alpha^3 = 1, \gamma, \alpha\gamma$ ($= \phi$) and $\alpha^2\gamma$ ($= \delta$). Thus $\langle X \rangle$ contains all six elements of G and so $G = \langle X \rangle$.

Taking $X = \{\alpha\}$, we see that $\langle X \rangle$ must contain at least the three elements $1, \alpha, \alpha^2$. However, we can check that these three elements do form a subgroup of G and so this subgroup is $\langle X \rangle$.

Example 4.10 A useful way to specify a group G is via *generators* and *relations*, giving a *presentation* for G. In this situation, we are given a list of generators such that the group consists of products of powers of these generators, with identifications deduced from the given list of relations. For example, the presentation $\langle a : a^n = 1 \rangle$ defines the cyclic group with n elements.

As another example, let n be a positive integer, and

$$G = \langle a, b : b^2 = 1 = a^n \text{ and } ab = ba^{-1} \rangle.$$

Clearly, if G is to be a group, it must contain the n powers of a, together with b and the products such as ba^i ($0 \leq i \leq n-1$). It should also contain products like $a^i b$. However, the relation $ab = ba^{-1}$ enables one to prove that $a^i b = ba^{-i}$ by induction on n. The anchor step here is trivial, since it is the given relation. Suppose, therefore, that $a^i b = ba^{-i}$. Then

$$a^{i+1}b = aa^i b = a(ba^{-i}) = ba^{-1}a^{-i} = ba^{-i-1},$$

as required.

To show that the elements $\{a^i, ba^j : 0 \leq i, j, \leq n-1\}$ are closed under multiplication, there are then four cases to consider.

(1) The product of elements of the form a^i and a^j is a^{i+j} and so is in G (after replacing a^{i+j} by a^{i+j-n} if $(i+j) \geq n$);
(2) the product of elements of the form ba^i and a^j is ba^{i+j} and so is in G (again after replacing a^{i+j} by a^{i+j-n} if $(i+j) \geq n$);
(3) the product of elements of the form a^i and ba^j is ba^{-i+j} and so is in G (after replacing a^{-i+j} by a^{n-i+j} if $i > j$) ; and finally

(4) the product of elements of the form ba^i and ba^j is

$$ba^iba^j = bba^{-i}a^j = a^{-i+j}$$

since $b^2 = 1$. It follows that this product is also in G (again after replacing a^{-i+j} by a^{n-i+j} if $i > j$).

This shows that the elements $\{a^i, ba^j : 0 \le i, j, \le n-1\}$ form a group.

There is, however, a problem to consider. How do we know that making more uses of the relations will not lead to unexpected conclusions such as $a = 1$? The group $\langle w : w^2 = 1, w^3 = 1\rangle$ has only one element because the relations mean that $1 = w^3 = w^2w = w$. One way to show that our group really does have $2n$ distinct elements is to find a 'model' for it. This can be done in this case, when $n > 2$, by giving an interpretation of the group as the symmetry group of a regular n-sided polygon, where a denotes anticlockwise rotation through $360/n$ degrees and b denotes a particular reflection. It can easily be checked that these symmetries satisfy the relations we have given for G. Since the symmetry group has n reflections and n rotations, it can be seen that the group with presentation given above has $2n$ elements. This group is known as the *dihedral group* $D(n)$.

We shall consider further examples of group presentations later and discuss the question of when presentations lead to unexpected results.

In the final part of this chapter, we consider properties of cyclic groups.

Proposition 4.11 *Let G be a group and g be any element of G. If g has infinite order, then $\langle g\rangle$ is an infinite cyclic group. If g has order n, then $\langle g\rangle$ has n elements.*

Proof Suppose that g has infinite order. Since $\langle g\rangle$ consists of all possible powers of g, this could only have a finite number of elements if some of these were equal. Then if $g^i = g^j$ with $i < j$, we would obtain

$$g^{j-i} = g^{i-i} = g^0 = 1,$$

so g would have finite order. Since this is contrary to hypothesis, we conclude that $\langle g\rangle$ is an infinite group.

If now g has finite order n, it follows by Proposition 3.10 that there are precisely n different elements of G of the form g^i, and so the group $\langle g\rangle$ has n elements. □

Proposition 4.12 *A cyclic group is abelian.*

Proof Let G be generated by an element g, so that each element of G is of the form g^i for some i. Thus if x and y are in G, $x = g^i$ and $y = g^j$ for some i and j. Then

$$\begin{aligned} xy &= g^i g^j = g^{i+j} \quad \text{by Proposition 3.8} \\ &= g^{j+i} = g^j g^i = yx, \end{aligned}$$

so G is abelian. □

Proposition 4.13 *A subgroup of a cyclic group is cyclic.*

Proof Let G be cyclic and generated by g, say, and H be a subgroup of G. Since H is a subset of G, all the elements of H are of the form g^k for some k. If $H = \{1\}$, then H is cyclic. We therefore suppose that H is not the identity subgroup and choose the element of H for which k is positive and as small as possible. Now let g^s be any other element of H. Use the division algorithm for integers to write s in the form $qk + r$ with $0 \leq r < k$. Then g^s and $(g^k)^{-1} = g^{-k}$ are in H, so since H is a subgroup, $g^s(g^{-k})^q = g^{s-kq} = g^r$ is also in H. This contradicts the definition of k, unless $r = 0$, in which case k divides s. We have shown that each element of H is of the form $(g^k)^q$ for some integer q, so H is cyclic and generated by g^k. □

We conclude this chapter with two basic properties of finite cyclic groups.

Proposition 4.14 *Let G be a finite cyclic group of order n. For any divisor d of n, the group G has precisely one subgroup G_d with d elements. The number of group elements x satisfying the equation $x^d = 1$ is d and these are precisely the elements of the subgroup G_d.*

Proof Let g be a generator of G, so that g has order n. It is clear that $g^{n/d}$ has d distinct powers, so $\langle g^{n/d} \rangle$ has d elements. If H were another subgroup of order d, by Proposition 4.13, H would be cyclic and generated by an element y, say. Then $y = x^r$ for some r, and since y has order d, we see that $x^{rd} = 1$. Then, by Proposition 3.10, n divides rd. Thus $kn = k(n/d)d = rd$ for some k. It follows that $r = k(n/d)$ so that n/d divides r. This means that y is a power of $g^{n/d}$, so that H is a subgroup of $\langle g^{n/d} \rangle$. Since these two subgroups have order d, they must be equal.

The above argument shows that x satisfies $x^d = 1$ if and only if x is a power of $g^{n/d}$. It follows that there are d solutions to this equation. □

Proposition 4.15 *Let G be a cyclic group with n elements, generated by an element g. Let $a = g^i$. Then a generates a cyclic subgroup H of G of order n/d, where d is the greatest common divisor of n and i.*

Proof It follows by Proposition 4.13 that a generates a cyclic subgroup H of G. We therefore only need to show that H has n/d elements. By Proposition 4.11, H has as many elements as the smallest power of a such that $a^k = 1$. Now $a = g^i$ and so $a^k = 1$ if and only if $(g^i)^k = 1$, or if and only if n divides ik. We may calculate the smallest integer k such that n divides ik as follows. Since d is the greatest common divisor of n and i, it follows by Theroem A.2 from Appendix A that there exist integers u and v such that

$$d = un + vi.$$

We may then write

$$1 = u(n/d) + v(i/d)$$

since d divides both n and i. This equation shows that n/d and i/d are relatively prime, because any integer which divides both of them must also divide 1. We wish to find the smallest integer k such that

$$\frac{ik}{n} = \frac{k(i/d)}{n/d}$$

is an integer. Since n/d and i/d are relatively prime, we conclude that n/d divides k, so that the smallest such integer is n/d. Thus H has n/d elements. □

Summary for Chapter 4

In this chapter, the important concept of a *subgroup* of a group G is introduced. A subgroup is defined to be a non-empty subset of G which is itself a group under the same operation as that of G. However, we soon obtain alternative criteria to enable one to check whether a given subset is a subgroup or not in Proposition 4.2. The most important construction discussed in this chapter is that of the subgroup *generated* by a subset X of a group G. This is, by definition, the intersection of all subgroups of G containing the set X. The most important special case of this occurs when the set X consists of a single element. In this case, the group $\langle X \rangle$ is said to be cyclic generated by X. Several further examples are given in the text, and these lead us to an informal discussion of the idea of a *presentation* of a group G in terms of generators and relations. The chapter ends with a discussion of the basic properties of cyclic groups.

Exercises 4

1. Which of the following sets H are subgroups of the given group G?
   (a) G is the set of integers under addition, H is the set of even integers;
   (b) $G = S(3), H = \{1, (12), (23), (13)\}$;
   (c) $G = GL(2, \mathbf{R})$, H is the set of matrices of the form $\begin{pmatrix} 1 & a \\ 0 & 1 \end{pmatrix}$, where a is any real number.

2. Give an example of a group G with subgroups H and K such that $H \cup K$ is *not* a subgroup of G.

3. Let G be the group in Question 2 of Exercises 1. Find the number of elements in $\langle A, D \rangle$. Is $\langle A, C \rangle$ cyclic? Write down the multiplication table for $\langle B, F \rangle$.

4. Let G be the group with presentation

$$\langle x, y : x^4 = 1,\ x^2 = y^2,\ xy = yx^{-1} \rangle.$$

Decide how many elements are in G and determine its multiplication table.

5

Cosets and Lagrange's Theorem

In this Chapter, Lagrange's Theorem will be proved. This important result states that the number of elements in a subgroup of a finite group G divides the number of elements of G. The starting point for a proof of Lagrange's Theorem is the idea of a coset.

Definition 5.1 *Let H be a subgroup of a group G and g be any element of G. The* left coset *gH is defined to be the set of group elements of the form gh, as h ranges over the elements of H:*

$$gH = \{gh : h \in H\}.$$

Remark 1 The coset $1H$ is simply the subgroup H. In fact for any h in H, since H is closed under multiplication, the coset hH is equal to H.

Remark 2 The element g is in the coset gH since $g = g1$.

Example 5.2 Let G be the dihedral group $D(3)$ of order 6, the group of symmetries of an equilateral triangle. Thus, as discussed in Chapter 4, G has presentation

$$G = \langle a, b : b^2 = 1 = a^3, ab = ba^{-1} \rangle.$$

Since b has order 2 the subgroup $H = \langle b \rangle$ has two elements. The six elements of G are $1, a, a^2, b, ba, ba^2$. We now calculate the six possible left cosets of H in G:

$$1H = 1\{1, b\} = \{1, b\} = H;$$

$$bH = b\{1, b\} = \{b, 1\} = H;$$

$$aH = a\{1, b\} = \{a, ab\} = \{a, ba^2\};$$

$$a^2H = a^2\{1, b\} \;=\; \{a^2, a^2b\} = \{a^2, ba\};$$

$$baH = ba\{1, b\} \;=\; \{ba, bab\} = \{ba, a^2\};$$

$$ba^2H = ba^2\{1, b\} \;=\; \{ba^2, ba^2b\} = \{ab, a\}.$$

Thus there are actually only three distinct left cosets, since the other three are equal to these:

$$1H = bH;\; aH = ba^2H;\; a^2H = \; baH.$$

As we shall see in the next three results, this is typical of the general situation.

Proposition 5.3 *Let H be a subgroup of G. Then the relation R on G defined by*

$$xRy \text{ if and only if } x^{-1}y \in H$$

is an equivalence relation.

Proof We check the three requirements on R, as given in the definition of an equivalence relation in Definition 2.24.

(1) Since H is a subgroup, $x^{-1}x = 1_G$ is in H, so xRx showing that R is reflexive.

(2) If xRy then $x^{-1}y \in H$. Since H is a subgroup,

$$y^{-1}x = (x^{-1}y)^{-1} \in H$$

so that yRx, showing that R is symmetric.

(3) If xRy and yRz, then $x^{-1}y$ and $y^{-1}z$ are both in H, so

$$(x^{-1}y)(y^{-1}z) = x^{-1}z \in H,$$

so that xRz. This checks that R is transitive. □

Proposition 5.4 *Let H be a subgroup of G and R be the equivalence relation of Proposition 5.3. Then the equivalence class of an element g of G is the left coset gH.*

Proof We first show that the equivalence class of g under R is contained in gH. This follows because if x is equivalent to g then $g^{-1}x = h$ for some $h \in$H. Thus $x = gh$ and so $x \in gH$.

Conversely, we show that every element of gH is contained in the equivalence class of g. Let gh be an element of gH, then $g^{-1}(gh) = h \in H$, so that $gR(gh)$, as required. □

Corollary 5.5 *Let H be a subgroup of a group G. Then two left cosets xH and yH of H in G are either equal or disjoint. In fact, $xH = yH$ if and only if $y^{-1}x \in H$. If xH is different from yH, then $xH \cap yH$ is the empty set. Each element of G is in some left coset of H in G.*

Proof This follows from the proposition together with the fact that the equivalence classes partition G, by Proposition 2.29. □

Example 5.6 We can illustrate these results using Example 5.2. We saw that

$$1H = bH; \quad aH = ba^2H; \quad a^2H = baH.$$

These equalities can now be seen as consequences of the facts that

$$b^{-1}1 = b^{-1} = b \in H;$$

$$(ba^2)^{-1}a = a^{-2}b^{-1}a = aba = b \in H; \text{ and}$$

$$(ba)^{-1}a^2 = a^{-1}b^{-1}a^2 = a^2ba^2 = b \in H.$$

Example 5.7 Let G be the additive group of integers $\mathbf{Z}$ and let n be any positive integer. The set $n\mathbf{Z}$ of integer multiples of n is easily seen to be a subgroup of $\mathbf{Z}$. The operation is addition so the identity element of $n\mathbf{Z}$ is 0. Choose any integer k and consider the left coset of $n\mathbf{Z}$ in $\mathbf{Z}$ containing k. Since the operation is addition, this is denoted by $k + n\mathbf{Z}$, and consists of integers of the form $k + n\ell$ as ℓ ranges over the set of integers. Another way to say this is that the left coset is the congruence class of k modulo n as considered in Chapter 2.

Another feature of these examples which holds in general is that each left coset of H in G has the same number of elements.

Proposition 5.8 *Let H be a subgroup of a group G. For any element g in G, there is a bijection α between H and gH.*

Proof We define a map α from H to gH by the rule $\alpha(h) = gh$ for all h in H. To show α is a bijection, we show it is injective and surjective.

To show that α is injective: suppose that x and y are elements of H and $\alpha(x) = \alpha(y)$, so that $gx = gy$. Then multiplying both sides of this equation on the left by g^{-1}, we deduce that $x = y$.

To show that α is surjective: a general element of gH is of the form gh for h in H, so it has the form $\alpha(h)$. □

Notation For any set X, the cardinality of X is denoted by $|X|$. When X is a finite group G, we often refer to $|G|$ as the *order of G*. In this notation, Proposition 5.8 shows that if H is a subgroup of a group G, then for any g in $G, |H| = |gH|$.

We now come to the main result of this section.

Theorem 5.9 (Lagrange) *Let G be a group with a finite number of elements and let H be a subgroup of G. Then the number r of distinct left cosets of H in G is equal to $|G|/|H|$. In particular, both $|H|$ and r divide $|G|$.*

Proof The equivalence classes under the relation R partition G. Each equivalence class is a left coset and each left coset has the same number of elements as H. It follows that if r is the number of distinct left cosets then $|G| = r|H|$. □

Definition 5.10 *The number of distinct left cosets of a subgroup H in a group G is the* index *of H in G. It is usually denoted by $|G : H|$.*

Example 5.11 As a simple application of this result, we determine all the subgroups of the group $G = C_2 \times C_2$. The four elements of G can be written in the form $(1,1), (x,1), (1,y)$, and (x,y) (where $x^2 = 1 = y^2$). Then the group itself and the subgroup consisting of the identity element are subgroups with 4 and 1 elements, respectively. It is also clear that the only subgroup with 4 elements is the group itself, and that the only subgroup with one element is the subgroup $\{(1,1)\}$. By Lagrange's Theorem, the order of any other subgroup divides 4 and so must be 2. A subgroup with two elements consists of the identity element together with an element which must have order 2. Thus there are three subgroups with two elements:

$$\{(1,1),(x,1)\}, \{(1,1),(1,y)\}, \text{ and } \{(1,1),(x,y)\},$$

and so there are five subgroups of G in all.

The word 'order' has now been used in two senses: the order of a group G (the number of elements in G), and the order of an element g in a group G

(the smallest positive integer k such that $g^k = 1$). It follows by Proposition 4.11 that the order of the element g is equal to the order of the subgroup $\langle g \rangle$.

Corollary 5.12 *Let g be an element of a finite group G. Then the order of g divides the order of G.*

Proof Apply Lagrange's Theorem to the subgroup $\langle g \rangle$ of G, noting that, by Proposition 4.11, the number of elements in this subgroup is equal to the order of g. □

Another important consequence of Lagrange's Theorem is the following:

Corollary 5.13 *Let H and K be subgroups of a finite group G with H a subgroup of K. Then*

$$|G : H| = |G : K|\ |K : H|.$$

Proof As we have seen,

$$|G : H| = |G|/|H| = (|G|/|K|)(|K|/|H|) = |G : K||K : H|,$$

as required. □

The discussion in this chapter has so far only concerned left cosets. Given a subgroup H of G and an element g of G, we can also define the right coset $Hg = \{hg : h \in H\}$. The early results in this chapter may be proved for right cosets, with minor modifications:

the relation xRy if and only if $xy^{-1} \in H$ is an equivalence relation;
the equivalence class of g under this relation is the right coset Hg;
two right cosets are either equal or disjoint; and
H and Hx have the same number of elements.

Moreover, there is a basic correspondence between left and right cosets.

Proposition 5.14 *Let H be a subgroup of a group G. The map α defined by $\alpha(gH) = Hg^{-1}$ is a bijection between the set of distinct left cosets of H in G and the set of distinct right cosets of H in G. In particular, the number of distinct left cosets of H in G is equal to the number of distinct right cosets of H in G.*

Proof To show that α is injective, suppose that $\alpha(xH) = \alpha(yH)$ for some $x, y \in G$. Then $Hx^{-1} = Hy^{-1}$, so $x^{-1} = 1x^{-1}$ is equal to hy^{-1} for some h in H. Thus $h = x^{-1}y$, and so $h^{-1} = y^{-1}x$. Since H is a subgroup, h^{-1} is in H, and so the left cosets xH and yH are equal by Corollary 5.5. This shows that α is injective. To show that α is surjective, let Hx be a right coset of H in G. Since x is $(x^{-1})^{-1}$, Hx is $\alpha(x^{-1}H)$ and so α is surjective, as required. $\square$

Example 5.15 Let us return to our earlier example and take

$$G = \langle b, a : b^2 = 1 = a^3, ab = ba^{-1}\rangle \text{ with } H = \langle b\rangle.$$

The distinct right cosets of H in G are

$$H = H1 = Hb;\ Ha = \{a, ba\} = Hba \text{ and } Ha^2 = \{a^2, ba^2\} = Hba^2.$$

Note that the elements in the right cosets are not the same elements as those in the left cosets: $\{a, ba\}$ is a right coset but not a left coset. The map α in Proposition 5.14 is just a correspondence between the left and right cosets which shows that there are an equal number of each. The elements in these cosets may be quite different.

The next topic is products of subgroups.

Definition 5.16 *Given two subsets A and B of a group G, we define the subset AB to be the set of all group elements of the form ab as a ranges over A and b ranges over B:*

$$AB = \{ab : a \in A,\ b \in B\}.$$

Proposition 5.17 *Let A and B be subgroups of a group G. Then AB is a subgroup of G if and only if $AB = BA$. We say that A and B are* permutable *in this case.*

Proof Suppose that AB is a subgroup of G, and let ab be in AB. Then ab is the inverse of an element c, say, of AB. We can write c as a_1b_1 for some a_1 in A and b_1 in B. Thus

$$ab = c^{-1} = (a_1b_1)^{-1} = b_1^{-1}a_1^{-1},$$

and so ab is in BA. This shows that $AB \subseteq BA$. To show the reverse inclusion, let $x \in BA$. Thus, $x = ba$ for some $a \in A$ and $b \in B$. Then $ba = (a^{-1}b^{-1})^{-1} \in AB$ since AB is a subgroup. It follows that $BA \subseteq AB$, as required.

Conversely, suppose that $AB = BA$. To show that AB is a subgroup of G, the subgroup requirements of Proposition 4.2(2) will be checked. Thus 1 is in AB since 1 is in A and in B. If a_1, a_2 are in A and b_1, b_2 are in B, then

$$(a_1b_1)(a_2b_2) = a_1(b_1a_2)b_2.$$

Since $AB = BA$, we see that $b_1a_2 = ab$ for some a in A and b in B. Thus

$$(a_1b_1)(a_2b_2) = (a_1a)(bb_2).$$

Since A and B are subgroups, it now follows easily that $(a_1b_1)(a_2b_2)$ is in AB, thus checking closure. Finally, if ab is in AB, then $(ab)^{-1} = b^{-1}a^{-1}$, so $(ab)^{-1}$is in $BA = AB$, and AB is closed under inverses. Thus AB is a subgroup of G. □

Proposition 5.18 *Let A and B be finite subgroups of a group G. Then*

$$|AB| = \frac{|A|\,|B|}{|A \cap B|}.$$

Proof Let

$$x_1(A \cap B),\ x_2(A \cap B), \ldots, x_k(A \cap B)$$

be the distinct left cosets of $A \cap B$ in A. Thus each element of A is in $x_i(A \cap B)$ for some $1 \leq i \leq k$, and also if $i \neq j$ then $x_j x_i^{-1}$ is not an element of $A \cap B$. Let ab be an arbitrary element of AB. We can then write a in the form $x_i g$ for some $1 \leq i \leq k$ and some g in $A \cap B$. Thus $ab = x_i(gb)$. Since g and b are in B, this shows that ab is in the coset x_iB. Furthermore, the cosets x_iB $(i = 1, \ldots, k)$ are disjoint: if not, $x_iB = x_jB$ for some i and j, by Corollary 5.5. Thus $x_j^{-1}x_i$ is an element of B, by Proposition 5.4. Since x_i and x_j are in the subgroup A, $x_j^{-1}x_i$ is in $A \cap B$, contrary to definition. We have shown that

$$r = |A|/|A \cap B| = |AB|/|B|,$$

and so

$$|AB| = \frac{|A|\,|B|}{|A \cap B|},$$

as required. □

We conclude this section with some further classification results.

Proposition 5.19 *Let p be a prime integer and G be a group with p elements. Then G is cyclic.*

Proof Suppose that G is a group with p elements and let g be a non-identity element of G. By Corollary 5.12, the order of g divides p, the order of G. Since p is a prime, the only divisors of p are 1 and p. The order of g is not 1 since g is a non-identity element, so the order of g is p. It follows by Proposition 4.11 that the subgroup $\langle g \rangle$ has p elements and so is the whole of G. Thus G is cyclic. □

This result immediately yields one of the results discussed in Chapter three, namely that groups of order three are cyclic. We now consider how to obtain the list of groups with six elements. This is another case where the method to be used now will be simplified later on when we have developed more theory.

Groups with six elements

We have already seen in Example 2.19 that a group with six elements exists. Let G be any group of order 6 (not necessarily the group of Example 2.19). By Corollary 5.12, the possible orders of elements of G are 1, 2, 3 or 6. If G has an element of order 6, then G is cyclic by Proposition 4.11. Thus a non-cyclic group with 6 elements has an element of order 1 (the identity) and each other element has order 2 or order 3. We show that if G is non-cyclic then G has an element of order 3, by showing that not every non-identity element can have order 2. If this were the case, take a and b to be distinct elements of order 2, then ab would have order 2. Then the inverse of ab would be ab:

$$ab = (ab)^{-1} = b^{-1}a^{-1} = ba.$$

It would then follow that the four elements $\{1, a, b, ab\}$ are a subgroup of G, which is impossible by Lagrange's Theorem. Thus G has an element a, say, of order 3.

We next show that G has an element of order 2. If not, each non-identity element has order 3. This would mean that each non-idenitity element could be paired with its inverse to give an even number of non-identity elements, which is impossible. It therefore follows that there must be an element of order 2.

We now have names for the six elements of $G : 1, b, a, a^2, ba$ and ba^2. Closure implies that ab is in G and we can easily see that the only candidates for ab are ba or $ba^2 = ba^{-1}$. In the first case, it may be checked that there are six distinct powers of ba, so that G is cyclic generated by ba. In the second case G has presentation

$$\langle b, a : b^2 = 1 = a^3, ab = ba^{-1} \rangle,$$

so G is isomorphic to the dihedral group $D(3)$.

Remark The bulk of the above discussion was taken up by proving that G has an element of order 2 and also an element of order 3. We shall see later that both these facts are consequences of the Sylow Theorems, to be discussed in Chapter 11, so that the methods just employed are only of temporary use.

Finally, we discuss one more example briefly, namely the classification of groups with eight elements. Most details will be left to the reader. Since we know that groups with 2, 3, 5 or 7 elements are cyclic (by Proposition 5.19), and groups of order 4 were classified at the end of Chapter 3, we shall know all groups with 8, or fewer, elements.

Groups with eight elements

The possible orders of elements in such a group are 1, 2, 4 or 8. If G has an element of order 8, then G is cyclic. If every element has order 1 or 2, then G is abelian (Question 1 of Exercises 3). In this case, it follows easily that G is isomorphic to $C_2 \times C_2 \times C_2$. We may therefore suppose that G has at least one element y, say, of order 4. Let x be any element not in $\langle y \rangle$, then the elements of G are $1, y, y^2, y^3, x, xy, xy^2$ and xy^3. Consider x^2: it cannot equal x, xy, xy^2 or xy^3; if it were equal to y or y^3 then x would have order 8 and G would be cyclic, so we may consider two cases:

(a) Suppose that $x^2 = 1$. Since yx is an element of G, it is one of the elements $1, y, y^2, y^3, x, xy, xy^2$ or xy^3. We can quickly see that the only possibilities that do not lead to an immediate contradiction are that yx is either xy or xy^3. (If $yx = xy^2, x^{-1}yx = y^2$ and squaring both sides would give that $y^2 = 1$). If $xy = yx$, it can be checked that G is abelian, and G is isomorphic to $C_4 \times C_2$. We obtain two possible presentations:

$$\langle x, y : \ y^4 = 1 = x^2 \text{ and } yx = xy \rangle \text{ and}$$

$$\langle x, y : \ y^4 = 1 = x^2 \text{ and } yx = xy^{-1} \rangle.$$

(b) Suppose that $x^2 = y^2$. As in (a), we see only two possibilities for yx, namely xy or xy^3. The first possibility gives a group with generators x, y and relations $y^4 = 1$, $xy = yx$ and $x^2 = y^2$. Writing $b = y$ and $a = yx^{-1}$, we see that b has order 4 and a has order 2, and since G is abelian in this case, G is isomorphic to $C_4 \times C_2$. The second possibility gives the quaternion group Q_8 with presentation

$$\langle x, y : y^4 = 1, y^2 = x^2 \text{ and } yx = xy^{-1} \rangle.$$

To demonstrate that a group satisfying these relations really exists, we define 2×2 matrices as follows:

$$X = \begin{pmatrix} 0 & 1 \\ -1 & 0 \end{pmatrix} \text{ and } Y = \begin{pmatrix} i & 0 \\ 0 & -i \end{pmatrix},$$

where i denotes a complex fourth root of unity. It is a simple matter to check that X and Y satisfy the given relations and that the eight matrices

$$I,\ X,\ X^2,\ X^3,\ Y,\ YX,\ YX^2,\ YX^3$$

form a group.

We have therefore seen that there are five groups with eight elements: the three abelian groups $C_8, C_4 \times C_2$ and $C_2 \times C_2 \times C_2$; the dihedral group $D(4)$ and the group Q_8. It is an easy calculation to see that the number of elements of order 2 in each of these five groups is 1, 3, 7, 5 and 1, respectively. Since isomorphic groups have the same number of elements of any order, and an abelian group cannot be isomorphic to a non-abelian group, we see that there are precisely five isomorphism types of groups of order 8.

Summary for Chapter 5

The main result in this chapter is Lagrange's Theorem, which states that the number of elements in any subgroup of a finite group G divides $|G|$. The key idea in the proof of Lagrange's Theorem is that of a (left) coset of a subgroup H of G. This is a subset of G of elements of the form gh as h ranges over the elements of H. It is shown that two left cosets are either equal or disjoint. The number of distinct left cosets of H in G is the *index* of H in G and is denoted by $|G : H|$. It follows from the proof of Lagrange's Theorem that $|G| = |G : H||H|$, so that the index of H in G also divides $|G|$. There are a number of important corollaries of this basic result, including the fact that any group of prime order is cyclic. The chapter concludes with a discussion of two classification results. It is shown that any group of order six is either cyclic or isomorphic to $D(3)$. Also, any group of order eight is either cyclic or isomorphic to one of $C_4 \times C_2$, $C_2 \times C_2 \times C_2$, $D(4)$, or the quaternion group Q_8.

The result we refer to as Lagrange's Theorem is actually due to Jordan. Lagrange showed that (in modern terminology) any subgroup of the symmetric group $S(n)$ has order dividing $n!$. However, the proof given by Lagrange used an idea equivalent to that of cosets and so can be generalised to arbitrary groups.

Exercises 5

1. Let G be the group of Question 2 in Exercises 1. Write down:

(a) the list of left cosets of the subgroup $\langle A \rangle$ in G;
(b) the list of left cosets of the subgroup $\langle B, F \rangle$ in G; and
(c) the list of left cosets and the list of right cosets for the subgroup $\{I, D\}$.

2. Show that if the left coset gH is a subgroup of G, then g is in H.

3. Show that if an element y of a group G is in the right coset Hx then $Hy = Hx$.

4. Show that two right cosets Hx, Hy of a subgroup H in a group G are equal if and only if yx^{-1} is an element of H.

5. Give an example of a group G with subgroups A and B such that AB is not a subgroup of G.

6. Let p be a prime number and G be a group with $p^a k$ elements, where a is a positive integer and p does not divide k. Suppose that P is a subgroup of G with p^a elements and Q is a subgroup of G with p^b elements, where $0 < b \leq a$. If Q is not a subgroup of P, show that PQ is not a subgroup of G.

6

Error-correcting codes

In this chapter, we introduce the elementary theory of error-correcting codes. In due course, the relevance of this topic to the theory introduced in Chapter 5 will be made clear.

The theory of codes was developed in order to ensure reliability of transmitted information. As an example, consider the ISBN (International Standard Book Number) of a published book. This number usually appears on the back of the book in the bottom right-hand corner. The ISBN consists of a nine-digit integer followed by a check digit which may be any one of the usual digits $0, 1, \ldots, 9$ or the symbol X (standing for 10). This final symbol may be calculated from the other nine as follows. Form an integer N by adding together the first digit, twice the second digit, three times the third and so on. The check digit is the remainder when N is divided by 11. For example, a book with first 9 digits 019853453 will have

$$N = 0 + 2 + 27 + 32 + 25 + 18 + 28 + 40 + 27 = 199$$

and so the check digit should be 1, giving ISBN 01953453 1. The point about such a number is that if it is inaccurately copied, and an error is made in any of the digits in the first nine locations (such as the last '5' being copied as a '3'), then the resulting number will not have '1' as its check digit. This is an example of an error-detecting code: the ISBN detects when a single error is made after transcribing the number.

In this chapter we shall explain methods which not only detect errors, but enable us to correct errors.

Definition 6.1 *Let p be a prime integer. Denote by $V(n, p)$ the set of all sequences of length n of elements from the set $\mathbf{Z}_p$ of congruence classes modulo p, so that $V(n, p)$ has p^n elements.*

We will usually omit square brackets and subscripts, so that $[i]_p$ will just be written as i. We also omit the commas and brackets commonly used to denote elements of vector spaces, so that $(1, 0, 1)$ will be written as 101. Thus $V(3, 2)$ consists of the eight sequences

$$000,\ 001,\ 010,\ 011,\ 100,\ 101,\ 110,\ 111,$$

while $V(2,3)$ consists of the nine sequences

$$00,\ 01,\ 02,\ 10,\ 11,\ 12,\ 20,\ 21,\ 22.$$

We add sequences by adding appropriate terms, remembering that we are adding congruence classes. Thus in $V(3,2)$

$$110 + 011 = 101;$$

while in $V(2,3)$

$$12 + 11 = 20.$$

We can also multiply an element in $V(n,p)$ by a congruence class by multiplying each term in the sequence by the representative for the congruence class and reducing modulo p. For example, in the space $V(3,3)$ we see that $2(102) = 201$. In fact $V(n,p)$ is a vector space of dimension n over the field $\mathbf{Z}_p$.

Definition 6.2 *A* linear (n,k)-code *is any k-dimensional subspace C of the vector space $V(n,p)$. Thus C satisfies the following two conditions:*

(1) *the difference of any two elements of C is an element of C; and*
(2) *the product of any element of C with any element of $\mathbf{Z}_p$ is also an element of C.*

The elements of C are called codewords.

Remark A subspace of a vector space is necessarily non-empty, so condition (1) ensures that the zero element of the vector space is in the code C. It then follows by the additive version of Proposition 4.2(3) that C is a group under addition.

Example 6.3 Consider the four elements 000, 001, 010, 011 of $V(3,2)$. These are precisely the four sequences which start with a 0. It is clear that subtracting any two of these gives a sequence starting with 0, and so condition (1) is satisfied. Since 0 and 1 are the only elements of $\mathbf{Z}_2$, condition (2) is automatically satisfied. Thus the four elements form a linear (3,2)-code.

Example 6.4 The three elements $00, 11, 22$ of $V(2,3)$ form a linear (2,1)-code.

Definition 6.5 *Let v be any element of $V(n,p)$. The* weight *of v is the number of non-zero terms in the sequence v. If v and w are two elements in $V(n,p)$ the* distance *$d(v,w)$ is the number of places at which v and w differ.*

Example 6.6 In $V(4,3)$ the weight of 1201 is three, since there are three non-zero entries. The distance from 1201 to 2211 is two, since these two vectors differ in two places.

Proposition 6.7 *Let u, v and w be any elements of $V(n,p)$. Then*

(1) $d(u,v) \geq 0$ *with equality if and only if* $u = v$;
(2) $d(u,v) = d(v,u)$; *and*
(3) $d(u,v) + d(v,w) \geq d(u,w)$.

Proof (1) It follows directly from the definition that $d(u,v)$ is positive except when u and v do not differ anywhere.

(2) This is clearly true.

(3) In each location at which u and w differ, v cannot agree with both u and w. Thus every contribution to the value of $d(u,w)$ provides a contribution to either $d(u,v)$ or to $d(v,w)$. □

Definition 6.8 *Let C be a subspace of $V(n,p)$. The* minimum distance *d of C is the least distance between different codewords:*

$$d = min_{u,v}\{d(u,v)\}.$$

The next result shows that for a linear code, the minimum distance can be calculated from the codewords.

Proposition 6.9 *Let C be a linear (n,k)-code. Then the minimum distance of C is equal to the smallest possible weight of any non-zero codeword.*

Proof Let f be the smallest possible weight of any non-zero codeword, and let *0* denote the sequence consisting entirely of zeros. Suppose that w is a codeword of weight f. Then $d(w, 0)$ is f, so $f \geq d$. Now let u and v be a pair of codewords with $d(u,v) = d$. Since C is a linear code, the word $u - v$ is a codeword of weight d, so $d \leq f$. It then follows that $d = f$. □

The importance of the idea of the minimum distance lies in the following result.

Proposition 6.10 *Let C be a linear code with minimum distance d. Then C detects $d - 1$ or fewer errors, and d corrects e errors for any e with $2e + 1 \leq d$.*

Proof Let v be a vector which has distance f from a codeword c, where $f \leq d - 1$. We think of c as the transmitted word and v as the received word, so that there are f errors in transmission. Since d is the minimum distance for C, the received v cannot be a codeword. We express this by saying that the code C detects d or fewer errors.

Suppose now that v has distance e from a codeword c and also that $2e + 1 \leq d$. Then there can be no other codeword nearer to v: if c_1 was in C and $d(v, c_1) \leq e$, then by Proposition 6.7(3),

$$d(c, c_1) \leq d(c, v) + d(v, c_1) \leq e + e < d,$$

contrary to definition. Thus, there is a unique nearest codeword to v, and we say that C corrects e errors in this case. □

Before giving examples of the use of this proposition, we give a way to construct linear codes.

Definition 6.11 *Let n and k be any positive integers with $n > k$. Let p be a prime number. A* (standard) generator matrix *G over $\mathbf{Z}_p$ is a $k \times n$ matrix with entries in $\mathbf{Z}_p$, in which the first k columns form an identity $k \times k$ matrix. Given such a matrix, we obtain a linear code by regarding the rows as sequences and taking all possible linear combinations of these. Alternatively, we can consider the code as consisting of all sequences obtained from matrix multiplications of the form uG as u varies over all sequences of length k over $\mathbf{Z}_p$.*

Example 6.12 Consider the generator matrix over $\mathbf{Z}_2$

$$G = \begin{pmatrix} 1 & 0 & 1 \\ 0 & 1 & 1 \end{pmatrix}.$$

The corresponding code consists of the combinations of the rows and so has four elements: 000; 101; 011 and 110. The codewords can also be described as the vectors of the form uG, as u varies over the four vectors 00, 01, 10, 11. Every non-zero codeword has weight 2, so the code detects one error, but does not correct errors. For example, 111 is not a codeword (so is *detected*) but is of equal distance from the two codewords 101 and 011, so cannot be corrected.

Example 6.13 Another example of a binary code (code over $\mathbf{Z}_2$) is provided by the matrix

$$\begin{pmatrix} 1 & 0 & 0 & 1 & 1 & 0 \\ 0 & 1 & 0 & 1 & 0 & 1 \\ 0 & 0 & 1 & 0 & 1 & 1 \end{pmatrix}$$

There are 8 codewords obtained from the rows of this matrix:

000000; 100110; 010101; 110011; 001011; 101101; 011110; 111000.

There are four codewords of weight 3, three codewords of weight 4 and one of weight 0. The minimum distance of this code is therefore 3, so the code detects two errors and corrects one error. For example, 100111 lies at distance one from a unique codeword, 100110, and so there is a unique way to correct one error. The vector 100001, however has distance two from 000000 and 110011, so cannot be corrected.

Example 6.14 Consider the following generator matrix over $\mathbf{Z}_3$:

$$\begin{pmatrix} 1 & 0 & 2 & 1 \\ 0 & 1 & 1 & 2 \end{pmatrix}.$$

In this case, the code consists of the linear combinations of the rows of the matrix, including multiplication by 1 and 2 since $p = 3$. There are 9 codewords:

0000; 1021; 2012; 0112; 1100; 2121; 0221; 1212 and 2200.

Since there is a codeword of weight 2 this code detects one error. Note that the minimum distance is 2 despite the fact that each row of the generator matrix has weight 3.

Example 6.15 As a final example, consider the following important code over $\mathbf{Z}_3$:

$$\begin{pmatrix} 1 & 0 & 0 & 0 & 0 & 0 & 0 & 1 & 2 & 2 & 1 \\ 0 & 1 & 0 & 0 & 0 & 0 & 1 & 0 & 1 & 2 & 2 \\ 0 & 0 & 1 & 0 & 0 & 0 & 2 & 1 & 0 & 1 & 2 \\ 0 & 0 & 0 & 1 & 0 & 0 & 2 & 2 & 1 & 0 & 1 \\ 0 & 0 & 0 & 0 & 1 & 0 & 1 & 2 & 2 & 1 & 0 \\ 0 & 0 & 0 & 0 & 0 & 1 & 1 & 1 & 1 & 1 & 1 \end{pmatrix}.$$

It is clear that the minimum distance of this code is at most 5 since there is a row of the generator matrix of weight 5. It can be shown that the minimum distance is exactly 5, so that the code corrects two errors. This is the *Golay ternary code* and is one of the most important codes. More details on this may be found in Hill (1986), which gives an introduction to the whole subject of error-correcting codes.

We now consider the problem of decoding a linear (n, k)-code C. This is done by listing the left cosets of the subgroup C of $V(n, p)$ in a table known as the coset decoding table. The table is organised by writing the codewords as its first row with the zero codeword first. Each subsequent row is a left coset of C. The entries in the first column are the coset representatives, now called coset leaders. The algorithm for choosing the r–th coset leader is to choose any word of minimum weight not already included in the first $r-1$ rows. Then to decode a given vector, locate it in the table, and correct it to the codeword standing in the same column of the coset decoding table.

Example 6.16 Consider Example 6.13 above. There are eight codewords which form a subgroup C of the vector space $V(6, 2)$. Since V has $2^6 = 64$ elements, this subgroup has index $64/8 = 8$. To form a complete coset decoding table, we list the elements of C in a row. We then choose any element v_2 which is of smallest weight among those not in the first row and write this at the left-hand end of the second row. The second row is obtained by adding each element of C in turn to this. Thus the second row is just the coset of C with respect to v_2. Continue this process by choosing v_3 to be of smallest weight among the elements not in the first two rows, and so on. This process is not unique, but depends upon the choice of coset representatives. One example of these choices is given in the following table:

000000; 100110; 010101; 110011; 001011; 101101; 011110; 111000;
100000; 000110; 110101; 010011; 101011; 001101; 111110; 011000;
010000; 110110; 000101; 100011; 011011; 111101; 001110; 101000;
001000; 101110; 011101; 111011; 000011; 100101; 010110; 110000;
000100; 100010; 010001; 110111; 001111; 101001; 011010; 111100;
000010; 100100; 010111; 110001; 001001; 101111; 011100; 111010;
000001; 100111; 010100; 110010; 001010; 101100; 011111; 111001;
100001; 000111; 110100; 010010; 101010; 001100; 111111; 011001.

To decode any element v of $V(6, 2)$, we locate v in the table and then correct it to the element in the first row of the column containing v. Thus, to use the table to decode 011010, we need to locate it (it is in the fifth row and seventh column) and correct it to the element in the the first row and the same column, giving 011110.

Note that the coset representative for the last row is not easy to find. According to the algorithm, we need a word of weight 2 not in the first seven rows. The representative we chose, 100001, is not unique.

This is actually a somewhat cumbersome way to arrange the decoding, since an exhaustive search is required. The calculation can be made more systematic for codes given by (standard) generator matrices using (standard) parity check matrices.

Definition 6.17 *Let C be an (n,k)-linear code over $\mathbf{Z}_p$ defined using a $k \times n$ generator matrix G of the form*

$$G = \begin{pmatrix} 1 & 0 & 0 & \dots & 0 & \\ 0 & 1 & 0 & \dots & 0 & \\ \vdots & & & & & A \\ 0 & 0 & 0 & \dots & 1 & \end{pmatrix}$$

where A is any $k \times (n-k)$ matrix. The parity check matrix *associated with G is the $(n-k) \times n$ matrix*

$$\begin{pmatrix} & 1 & 0 & 0 & \dots & 0 \\ & 0 & 1 & 0 & \dots & 0 \\ -A^{\mathrm{T}} & \vdots & & & & \\ & 0 & 0 & 0 & \dots & \end{pmatrix}$$

Notation The generator matrix G above is often written, in block matrix form, as $G = (I_k \mid A)$. Similarly, the parity check matrix is written as $P = (-A^{\mathrm{T}} \mid I_{n-k})$.

Example 6.18 The parity check matrix of the generator matrix over $\mathbf{Z}_2$

$$\begin{pmatrix} 1 & 0 & 0 & 1 & 1 & 0 \\ 0 & 1 & 0 & 1 & 0 & 1 \\ 0 & 0 & 1 & 0 & 1 & 1 \end{pmatrix}$$

is the matrix

$$\begin{pmatrix} 1 & 1 & 0 & 1 & 0 & 0 \\ 1 & 0 & 1 & 0 & 1 & 0 \\ 0 & 1 & 1 & 0 & 0 & 1 \end{pmatrix}.$$

Definition 6.19 *Let C be a linear (n,k)-code with generator matrix G and associated parity check matrix P. For any v in $V(n,p)$, let v^{T} denote the transpose of v, the column vector obtained by writing the members of the sequence v vertically. Then the* syndrome *of v is the element of $V(n-k,p)$ given by Pv^{T}.*

Thus in the above example, the syndrome of 100000 is 110 and the syndrome of 110011 is 000.

Proposition 6.20 *Let C be a code with standard parity check matrix P. An element v in $V(n,p)$ is a codeword if and only if the syndrome of v is the zero sequence.*

Proof Let $G = (I \mid A)$ be the standard generator matrix for C. Thus an element c of $V(n,p)$ is a codeword if and only if c is of the form uG, where u is in $V(k,p)$. Hence, the element c is a codeword if and only if $c = uG = u(I \mid A) = (u \mid v)$, where $v = uA$, and $(U \mid v)$ denotes the *concatenation* of u with v, that is, the row n-vector whose first k columns are u and last $n-k$ columns are v. For any w in $V(n,p)$, we may write w as the concatenation $(u_1 \mid v_1)$.

The syndrome of w is then

$$\begin{aligned}(-A^{\mathrm{T}} \mid I)w^{\mathrm{T}} &= (-A^{\mathrm{T}} \mid I)(u_1^{\mathrm{T}} \mid v_1^{\mathrm{T}}) \\ &= -A^{\mathrm{T}}u_1^{\mathrm{T}} + v_1^{\mathrm{T}}.\end{aligned}$$

It follows that the syndrome of w is zero if and only if w is the concatenation of u_1 and v_1, where $v_1^{\mathrm{T}} = A^{\mathrm{T}}u_1^{\mathrm{T}} = (u_1A)^{\mathrm{T}}$. Thus the syndrome of w is zero if and only if w is a codeword. □

Corollary 6.21 *Let C be a code with parity check matrix P. Two elements in $V(n,p)$ have the same syndrome if and only if they are in the same coset of C.*

Proof Let u and v be in $V(n,p)$. Then u and v have the same syndrome

if and only if $Pu^{\mathrm{T}} = Pv^{\mathrm{T}}$

if and only if $P(u^{\mathrm{T}} - v^{\mathrm{T}}) = 0$

if and only if $P(u - v)^{\mathrm{T}} = 0$

if and only if $u - v$ is in C

if and only if u and v are in the same coset of C,

as claimed. □

Example 6.22 Thus we need not store the complete coset decoding table, but merely a table of two columns: the coset representatives and their syndromes. In our previous example in which P was

$$\begin{pmatrix} 1 & 1 & 0 & 1 & 0 & 0 \\ 1 & 0 & 1 & 0 & 1 & 0 \\ 0 & 1 & 1 & 0 & 0 & 1 \end{pmatrix},$$

this table would be

coset representatives	syndromes
000000	000
100000	110
010000	101
001000	011
000100	100
000010	010
000001	001
100001	111

Thus to decode a given vector such as 100111, calculate its syndrome to obtain 001. This is the syndrome for the seventh row, so this vector is not a codeword, but the word 100110 obtained by subtracting 000001 is a codeword.

Remark Another advantage of listing coset representatives together with syndromes is that it is then much easier to find any missing coset representatives, since each sequence in $V(n-k,p)$ occurs as a syndrome. Thus in Example 6.22, the syndrome for the last row must be 111 because the other seven sequences of length 3 have already been used as syndromes. This enables us to find a representative relatively easily (compared with searching through the first seven rows), by seeing how to combine known coset leaders and their syndromes to obtain 111. Thus other possibilities would be 000111, 010010 or 001100.

Summary for Chapter 6

In Chapter 6, the theory of error-correcting codes over $\mathbf{Z}_p$ is discussed. Attention is restricted to linear codes, since these can be defined using generator matrices. In the language of group theory, a linear code is an additive subgroup of a vector space $V(n,p)$ of n-tuples over $\mathbf{Z}_p$. The problem of decoding, or correcting, errors is shown to be equivalent to calculating a set of coset representatives for the code viewed as a subgroup in this way. The effectiveness of a given code in correcting or detecting errors is measured by its least distance d. Given any v in V, the weight of v is the number of non-zero entries in the sequence v. The minimun distance of a linear code C is the least weight of any non-zero codeword of C. For any e with $2e + 1 \leq d$, the code C can correct e errors.

The theory of error-correcting codes was originally developed by electrical engineers in order to ensure correct transmission of electrical signals in computers. The pioneers in the subject were Golay and Hamming. Since then the method has been used to return messages from voyages into space. For example the photographs returned from space missions to the planets are routinely corrected for errors.

Exercises 6

1. For any element x in $\mathbf{Z}_2$, let $\bar{x}$ denote $1 + x$, so that $\bar{x}$ is 0 when x is 1 and $\bar{x}$ is 1 when x is zero. Let C be the set of elements of $V(6, 2)$ of the form $xyz\bar{x}\bar{y}\bar{z}$. Write down the eight elements of C, and show that C is not a linear code. What is the minimum distance of C?

2. In each of the following cases, say how many errors the code with the given generator matrix G detects and how many errors the code corrects:

(a) the code over $\mathbf{Z}_2$ with $G = \begin{pmatrix} 1 & 0 & 0 & 0 & 1 \\ 0 & 1 & 0 & 1 & 0 \\ 0 & 0 & 1 & 1 & 1 \end{pmatrix}$;

(b) the code over $\mathbf{Z}_3$ with $G = \begin{pmatrix} 1 & 0 & 1 & 1 \\ 0 & 1 & 1 & 2 \end{pmatrix}$;

(c) the code over $\mathbf{Z}_5$ with $G = \begin{pmatrix} 1 & 0 & 0 & 2 & 1 \\ 0 & 1 & 0 & 1 & 3 \\ 0 & 0 & 1 & 4 & 1 \end{pmatrix}$.

3. Let C be a linear code over $\mathbf{Z}_2$. Let C^+ be the subset of C consisting of those elements of C with even weight. Show that C^+ is an (additive) subgroup of C. By considering the cosets of the subgroup C^+ in C, show that either $C^+ = C$ or C^+ contains half the elements of C.

4. Construct a complete coset decoding table for the code in Question 2(b) above.

5. Calculate the parity check matrix for the code over $\mathbf{Z}_2$ with generator matrix

$$\begin{pmatrix} 1 & 0 & 0 & 1 & 1 & 0 & 1 \\ 0 & 1 & 0 & 1 & 1 & 1 & 0 \\ 0 & 0 & 1 & 1 & 0 & 1 & 1 \end{pmatrix}$$

and use it to construct the two-column decoding table. Decode the following:

1100011 1011000 0101110 0110001 1010110.

7

Normal subgroups and quotient groups

In this chapter we introduce the idea of a normal subgroup and show how these subgroups are used to construct quotient groups.

Definition 7.1 *Let x be an element of a group G. Any element of G of the form gxg^{-1} is a* conjugate *of x. For any subgroup H of G, $gHg^{-1} = \{gxg^{-1} : x \in H\}$ is a* conjugate *of H.*

Proposition 7.2 *Let H be a subgroup of the group G. Then any conjugate gHg^{-1} is also a subgroup of G.*

Proof In order to show that gHg^{-1} is a subgroup of G, we verify condition 2 of Proposition 4.2. The identity element of G is in gHg^{-1} since $g1g^{-1} = 1 \in gHg^{-1}$. If $a = gxg^{-1}$ and $b = gyg^{-1}$ are in gHg^{-1}, then

$$ab = (gxg^{-1})(gyg^{-1}) = g(xy)g^{-1} \in gHg^{-1}.$$

Finally, for any $gxg^{-1} \in gHg^{-1}$, we see that

$$(gxg^{-1})^{-1} = (g^{-1})^{-1}x^{-1}g^{-1} = gx^{-1}g^{-1} \in gHg^{-1},$$

as required. □

Definition 7.3 *A subgroup N of a group G is* normal *if and only if, for all g in G, $gNg^{-1} \subseteq N$ (so that for all $x \in N$ and all $g \in G$, $gxg^{-1} \in N$).*

Remark In any group G, $\{1\}$ and G are normal subgroups.

Proposition 7.4 *The following conditions on a subgroup N of a group G are equivalent:*

(a) *N is a normal subgroup of G;*
(b) *$g^{-1}Ng \subseteq N$ for all g in G;*

(c) $g^{-1}Ng = N = gNg^{-1}$ *for all* g *in* G;
(d) $gN = Ng$ *for all* $g \in G$; *and*
(e) *every right coset of* N *in* G *is a left coset.*

Proof (a) $\Rightarrow$ (b) For any $y \in G$, since N is a normal subgroup of G, $yNy^{-1} \subseteq N$. Taking y to be g^{-1}, gives that $g^{-1}N(g^{-1})^{-1} = g^{-1}Ng \subseteq N$.

(b) $\Rightarrow$ (c) For any $g \in G$, by (b), $gNg^{-1} = (g^{-1})^{-1}Ng^{-1} \subseteq N$, so

$$gN = gNg^{-1}g \subseteq Ng.$$

It follows that $N = g^{-1}(gN)$ is contained in $g^{-1}Ng$. Thus $N = g^{-1}Ng$, and so $N = gNg^{-1}$.

(c) $\Rightarrow$ (d) If $g^{-1}Ng = N$ for all g in G, we see that

$$gN = g(g^{-1}Ng) = Ng \text{ for all } g \text{ in } G.$$

(d) $\Rightarrow$ (e) This implication is obvious.

(e) $\Rightarrow$(a) For any g in G, condition (e) gives that the right coset Ng is a left coset, say $Ng = g'N$, for some g'. Then since 1 is in $g^{-1}Ng = g^{-1}g'N$, we have $1 = g^{-1}g'n$ for some n in N. It follows that each element of N is in the coset $g^{-1}g'N = g^{-1}Ng$. We have shown that $N \subseteq g^{-1}Ng$ and it follows easily from this that $gNg^{-1} \subseteq N$, so that N is a normal subgroup of G. □

Example 7.5 If G is abelian, then for any subgroup N of G and any element g in G, $gNg^{-1} = N$ since $gng^{-1} = gg^{-1}n = n$. Thus every subgroup of an abelian group is normal.

Example 7.6 Suppose that H is a subgroup of G of index 2, so that H has two left (and two right) cosets in G. Then if g is in H, the left coset gH is equal to H as is the right coset Hg. For any g not in H, the left coset gH must equal $G\backslash H$, since there are only two left cosets and the left cosets partition G. However, the right coset Hg must also equal $G\backslash H$ for the same reason. Thus every left coset is a right coset and so by Proposition 7.4(e), H is a normal subgroup of G.

Example 7.7 Normality is not a transitive relation. In the group $D(4)$ of order 8, $N = \{1, b, a^2, ba^2\}$ is a subgroup of order 4, and so is normal by Example 7.6. Now consider the subgroup $K = \{1, b\}$ of N. Since N is abelian, K is a normal subgroup of N. However, K is not a normal subgroup of G: the left coset $aK = \{a, ab\} = \{a, ba^{-1}\}$, but the right coset $Ka = \{a, ba\}$, so that the intersection of aK with Ka has one element. It follows that Ka cannot be a left coset.

We now give some elementary properties of normal subgroups.

Proposition 7.8 *Let N be a normal subgroup of a group G and H be any subgroup of G. Then $H \cap N$ is a normal subgroup of H.*

Proof For any x in $H \cap N$ and any g in H, gxg^{-1} is in H since H is a subgroup of G. However, since x is in N and N is a normal subgroup of G, gxg^{-1} is also in N. Hence gxg^{-1} is in $H \cap N$, as required. □

Proposition 7.9 *Let I be any index set and $\{N_i : i \in I\}$ be a set of normal subgroups of the group G. Then $\langle N_i : i \in I\rangle$ and $\bigcap N_i$ are normal subgroups of G.*

Proof By definition, $\langle N_i\rangle$ is the intersection of all the subgroups of G which contain each N_i. For any such subgroup H, its conjugate gHg^{-1} contains each gN_ig^{-1}. Since each N_i is a normal subgroup of G, we deduce that the conjugate gHg^{-1} contains N_i. It follows that

$$\langle N_i : i \in I\rangle$$

is a normal subgroup of G.

Also, since $\bigcap N_i$ is a subgroup of G by Proposition 4.6, and

$$g\left(\bigcap N_i\right)g^{-1} = \bigcap \left(gN_ig^{-1}\right)$$

for all $g \in G$, we deduce that $\bigcap N_i$ is a normal subgroup of G. □

Proposition 7.10 *Let N be a normal subgroup of the group G and H be any subgroup of G. Then $\langle N, H\rangle = NH$, so that $HN = NH$.*

Proof It is clear from the closure requirement that $\langle N, H\rangle$ must contain NH. It will therefore be sufficient to show that NH is a subgroup of G, since the result will then follow by definition of $\langle N, H\rangle$. Clearly, 1 is in NH. Suppose that $n, n_1 \in N$ and $h, h_1 \in H$. Then

$$(nh)(n_1h_1) = n(hn_1h^{-1}h)h_1 = n(hn_1h^{-1})(hh_1).$$

Since N is a normal subgroup, hn_1h^{-1} is an element n_2, say, of N, and so

$$(nh)(n_1h_1) = (nn_2)(hh_1) \in NH.$$

Also

$$(nh)^{-1} = h^{-1}n^{-1} = (h^{-1}n^{-1}h)h^{-1} \in NH,$$

so NH is a subgroup, as required. The fact that $HN = NH$ then follows by Proposition 5.17. □

The importance of the normality condition on a subgroup lies in the ability to construct a *quotient group* from the left cosets (right cosets would work equally here since each left coset is a right coset).

Proposition 7.11 *Let N be a normal subgroup of the group G. The set of left cosets of N in G is a group denoted by G/N under the operation $(xN)(yN) = xyN$.*

Proof We can check that the set of left cosets satisfy the four group axioms under the defined operation, but that is not the critical part of our proof. The claim that G/N is a group contains the implicit assertion that the operation is well-defined in the following sense. The rule that $(xN)(yN) = xyN$ appears to depend on the choice of coset representatives. Since left cosets are equivalence classes, this is a situation similar to one we encountered in Chapter 2, when discussing addition of congruence classes. The coset xN is equal to uN whenever $u^{-1}x \in N$, and $yN = vN$ whenever $v^{-1}y \in N$. We must check that if $xN = uN$ and $yN = vN$, then the coset xyN is equal to the coset uvN. To do this, we must check that $(uv)^{-1}(xy) \in N$ (or in terms of equivalence relations, if xRu and yRv then $xyRuv$). This follows from the following calculation:

$$\begin{aligned}(uv)^{-1}(xy) &= v^{-1}u^{-1}xy \\ &= v^{-1}ny \quad \text{where } n = u^{-1}x \in N \\ &= v^{-1}y(y^{-1}ny).\end{aligned}$$

Since N is a normal subgroup, $y^{-1}ny$ is in N, and $v^{-1}y$ is an element of N, so $(uv)^{-1}(xy)$ is also in N, as required. Notice that this argument uses the fact that N is a normal subgroup of G. Having established that the operation is well-defined, one easily checks the group axioms: the identity for the quotient group G/N is the coset $1N = N$, and the inverse of gN is $g^{-1}N$. □

Example 7.12 As a first example, consider the additive group of integers, **Z**. Since this is abelian, every subgroup of **Z** is normal. Consider the subgroup $n\mathbf{Z}$ consisting of all multiples of n. Let us determine the quotient group $\mathbf{Z}/n\mathbf{Z}$.

The elements of this group are the cosets $x + n\mathbf{Z}$. The integers in a given coset $x + n\mathbf{Z}$ are precisely the integers in the congruence class of x modulo n, as discussed in Chapter 2. Since **Z** is a group under addition, the operation in the quotient group $\mathbf{Z}/n\mathbf{Z}$ is also denoted as addition, and is given by the rule:

$$(x + n\mathbf{Z}) + (y + n\mathbf{Z}) = (x + y + n\mathbf{Z}).$$

The question of 'well-definedness' was discussed in Chapter 2, and may be expressed in the form: if u is congruent to x modulo n and v is congruent to y modulo n, then $x + y$ is congruent to $u + v$ modulo n.

There are n distinct cosets of $n\mathbf{Z}$ in $\mathbf{Z}$:

$$0 + n\mathbf{Z},\ 1 + n\mathbf{Z}, \ldots, (n - 1) + n\mathbf{Z},$$

so the quotient group is an abelian group with n elements. This group is cyclic generated by $1 + n\mathbf{Z}$.

We have therefore seen that the quotient group $\mathbf{Z}/n\mathbf{Z}$ is in fact the group $\mathbf{Z}_n$.

Example 7.13 As a second example, let G be the dihedral group $D(4)$:

$$G = \langle a, b : \ a^4 = 1 = b^2 \text{ and } ab = ba^{-1} \rangle.$$

Let N be the subgroup generated by a^2, so that N consists of the two elements 1 and a^2. Then N is a normal subgroup of G. This may be seen in a variety of ways, one way being to enumerate the four left cosets of N in G as

$$1N = \{1, a^2\},\ bN = \{b, ba^2\},\ aN = \{a, a^3\} \text{ and } baN = \{ba, ba^3\}.$$

Since $ba^2 = a^2b$ and $ba^3 = a^2ba$, it is easy to see that each of these is a right coset:

$$1N = N1,\ bN = Nb,\ aN = Na \text{ and } baN = Nba.$$

We next construct the multiplication table for G/N and, in order to do this, give names to the cosets, say $E = N, A = aN, B = bN$ and $C = abN$. Then the table is completed by calculations such as

$$AB = (aN)(bN) = abN = C;$$

$$AC = (aN)(abN) = a^2bN = bN = B,$$

to give the following table:

	E	A	B	C
E	E	A	B	C
A	A	E	C	B
B	B	C	E	A
C	C	B	A	E

This group is therefore isomorphic to the direct product $C_2 \times C_2$.

Remark Another way to view this table is to start with the standard table for G:

	1	a	a^2	a^3	b	ba	ba^2	ba^3
1	1	a	a^2	a^3	b	ba	ba^2	ba^3
a	a	a^2	a^3	1	ba^3	b	ba	ba^2
a^2	a^2	a^3	1	a	ba^2	ba^3	b	ba
a^3	a^3	1	a	a^2	ba	ba^2	ba^3	b
b	b	ba	ba^2	ba^3	1	a	a^2	a^3
ba	ba	ba^2	ba^3	b	a^3	1	a	a^2
ba^2	ba^2	ba^3	b	ba	a^2	a^3	1	a
ba^3	ba^3	b	ba	ba^2	a	a^2	a^3	1

Reordering the elements of G with respect to the coset decomposition of the normal subgroup $\{1, a^2\}$ gives the table

	1	a^2	a	a^3	b	ba^2	ba	ba^3
1	1	a^2	a	a^3	b	ba^2	ba	ba^3
a^2	a^2	1	a^3	a	ba^2	b	ba^3	ba
a	a	a^3	a^2	1	ba^3	ba	b	ba^2
a^3	a^3	a	1	a^2	ba	ba^3	ba^2	b
b	b	ba^2	ba	ba^3	1	a^2	a	a^3
ba^2	ba^2	b	ba^3	ba	a^2	1	a^3	a
ba	ba	ba^3	ba^2	b	a^3	a	1	a^2
ba^3	ba^3	ba	b	ba^2	a	a^3	a^2	1

Each of the 2×2 blocks in this table has entries from precisely one of the four cosets of N in G. It is the fact that N is a normal subgroup of G which allows this to happen. This reordering of the elements of G can then be seen as an expanded version of the table for G/N.

In a similar way, the original table falls into four 4×4 blocks. Each of these blocks only contains entries from one of the subsets $\{1, a, a^2, a^3\}$ or $\{b, ba, ba^2, ba^3\}$. This can be interpreted as saying that G has a normal subgroup $\{1, a, a^2, a^3\}$ with quotient group cyclic of order 2.

We conclude this chapter with a discussion of the properties of an important bijection concerning subgroups of quotient groups.

Theorem 7.14 (The Correspondence Theorem) *Let N be a normal subgroup of a group G. Then every subgroup of the quotient group G/N is of the form H/N for some subgroup H of G with $N \leq H$.*

Conversely, if H is a subgroup of G containing N then H/N is a subgroup of G/N. The correspondence between subgroups of G/N and subgroups of G containing N is a bijection. This bijection maps normal subgroups of G/N onto normal subgroups of G which contain N.

Proof Let $H^\star$ be a subgroup of G/N, so that it consists of a certain set $\{hN\}$ of left cosets of N in G. We define the subset $\beta(H^\star)$ of G to be $\{g \in G : gN \in H^\star\}$. Then $\beta(H^\star)$ clearly contains N and is a subgroup of G:

$1 \in N$, so $1 \in \beta(H^\star)$;
if $x, y \in \beta(H^\star)$, then xN and $yN \in H^\star$, so $(xN)(yN) = xyN \in H^\star$, and so $xy \in \beta(H^\star)$; and
since $(xN)^{-1} = x^{-1}N$, it follows that $x^{-1} \in \beta(H^\star)$.

Thus $\beta(H^\star)$ is a subgroup of G containing N.

Conversely, if H is any subgroup of G containing N, let $\alpha(H)$ be the subset $\{hN : h \in H\}$ of G/N. It may be easily checked that $\alpha(H)$ is a subgroup of G/N.

To complete the proof, we show that the map α from the set X of subgroups of G containing N to the set Y of subgroups of G/N is a bijection. This will be done by checking that the map $\beta : Y \to X$ is the inverse of α. To do this, we show that the composite maps $\alpha \circ \beta$ and $\beta \circ \alpha$ are the identity. Suppose that $H \leq G$ and $N \leq H$, then

$$\beta \circ \alpha(H) = \beta(H/N) = \{g \in G : gN \in H/N\} = H.$$

Conversely, let $H^\star$ be a subgroup of G/N, then

$$\alpha \circ \beta(H^\star) = \alpha(\{g \in G : gN \in H^\star\}) = \{gN \in H^\star\} = H^\star.$$

This completes the proof that α is a bijection by Proposition 2.15.

Now let H be a normal subgroup of G containing N. We shall show that $\alpha(H)$ is a normal subgroup of G/N. This follows since

$$(gN)(hN)(gN)^{-1} = ghg^{-1}N \in H/N.$$

Conversely, if $H^\star$ is a normal subgroup of G/N, it may easily be checked that $\beta(H^\star) = \{g \in G : gN \in H^\star\}$ is a normal subgroup of G. □

Example 7.15 Returning to the example of the quotient group of the dihedral group G of order 8 by the subgroup N of order 2, we saw that this was isomorphic to the group $C_2 \times C_2$. There are five subgroups of this direct product: the group itself, the identity subgroup, and three subgroups with two elements: $\{I, A\}, \{I, B\}$ and $\{I, C\}$. This means that there are five subgroups of G containing N, these being G, N and the three subgroups

$$\{1, a, a^2, a^3\}, \{1, b, a^2, ba^2\} \text{ and } \{1, ba, a^2, ba^3\}.$$

Since the group $C_2 \times C_2$ is abelian, each of its subgroups is normal. It follows that not only G and N, but the other three subgroups of G containing N are normal subgroups of G. (This also follows since each of these three subgroups has order 4 and therefore has index 2.)

Proposition 7.16 *Let N be a normal subgroup of a group G and let A, B be subgroups of G with $N \leq A, B$. Then for the bijective correspondence α of Theorem 7.14,*

(i) $\alpha(A \cap B) = \alpha(A) \cap \alpha(B)$;
(ii) $\alpha(\langle A, B \rangle) = \langle \alpha(A), \alpha(B) \rangle$.

Proof (i) We have that

$$\alpha(A) \cap \alpha(B) = \{gN : g \in A \cap B\} = (A \cap B)/N = \alpha(A \cap B).$$

(ii) The result follows since

$$\begin{aligned} \alpha(\langle A, B \rangle) &= \{hN : h \in \langle A, B \rangle\} \\ &= \{hN \in \langle A/N, B/N \rangle\} \\ &= \langle \alpha(A), \alpha(B) \rangle, \end{aligned}$$

as required. □

Remark In the next chapter, we shall see how normal subgroups and quotient groups arise naturally from the study of maps between groups.

Summary for Chapter 7

This chapter introduces the important idea of a *normal subgroup* of a group G. A subgroup N is normal if for all $g \in G$ and all $x \in N$, $gxg^{-1} \in N$. Given a normal subgroup N of G, the set of left cosets of N in G forms a group, denoted by G/N. Examples of this construction are given in the text. The

Correspondence Theorem shows how to determine all the subgroups and also the normal subgroups of G/N.

Galois was aware of the importance of the concept of a normal subgroup. He discussed this in the letter he hastily wrote to his friend Auguste Chevalier on the night before his duel. The idea of a quotient group appeared implicitly in the work of Camille Jordan in the 1860s, but the modern formulation using cosets is due to Otto Hölder in 1889, relatively late in the history of group theory.

Exercises 7

1. Let H be any subgroup of a group G and let g be any element of G. Prove that gHg^{-1} is a subgroup of G.

2. List all the subgroups of the dihedral group $D(3)$, and determine which of these are normal.

3. Let G be the group Q_8 discussed during the classification of groups of order eight in Chapter 5. Let N be the subset $\{1, x^2\}$. Show that N is a subgroup of G. By listing cosets, show that N is a normal subgroup of G, and determine the multiplication table for G/N.

4. Let G be the dihedral group $D(4)$:

$$G = \langle b, a : b^2 = 1 = a^4 \text{ and } ab = ba^{-1} \rangle,$$

and H be the subset $\{1, b\}$. Prove that H is not a normal subgroup of G. Show that multiplication of the left cosets of H in G is not well-defined: there are elements x, y, u and v with $xH = uH, yH = vH$, but $xyH \neq uvH$.

5. For any group G, define the *centre of* G to be the set of all elements z which commute with every element g of G:

$$Z(G) = \{z \in G : zg = gz \text{ for all } g \text{ in } G\}.$$

Prove that $Z(G)$ is a normal abelian subgroup of G and determine the list of elements in $Z(G)$ when G is $D(3)$ and also when G is $D(4)$.

8

The Homomorphism Theorem

In this chapter, we investigate maps between groups which preserve multiplication. The connection between this idea and the quotient groups which we studied in the last chapter is provided by Theorem 8.13, known as the Homomorphism Theorem.

Definition 8.1 *Let G and H be groups. A* homomorphism *from G to H is a map $\phi : G \to H$ such that, for all x, y in G, $\phi(xy) = \phi(x)\phi(y)$.*

In terms of the multiplication tables of G and H, this definition says that the entries in the table of G are taken by ϕ to corresponding entries in the table of H.

Example 8.2 Let G be the group $GL(n, \mathbf{R})$ of invertible $n \times n$ matrices with real entries under matrix multiplication and H be the multiplicative group of non-zero real numbers. For X in G, define $\phi(X)$ to be the determinant of X. Since X is invertible, its determinant is non-zero. It is a standard property of determinants that

$$\det(XY) = \det(X)\det(Y).$$

It follows from this that ϕ is a homomorphism.

Example 8.3 We saw in Chapter 2 that the symmetric group $S(3)$ is isomorphic to the dihedral group $D(3)$. Isomorphism is a particular type of homomorphism, in that ϕ is also required to be bijective.

Example 8.4 Let G be the group C_3 of complex third roots of unity $1, \omega$, and ω^2, where $\omega = e^{2\pi i/3}$. We can define a map ϕ from G to itself by $\phi(1) = 1, \phi(\omega) = \omega^2$ and $\phi(\omega^2) = \omega$. It may be checked that this map is a homomorphism.

Example 8.5 We saw in Chapter 2 that the group $\mathbf{Z}_3$ of congruence classes modulo 3 is a group under addition. This is isomorphic to the group C_3 of Example 8.4 even though the operations are different. In the case when the operation in G is written as addition and the operation in H is written as multiplication, the homomorphism requirement becomes

$$\vartheta(g+h) = \vartheta(g)\vartheta(h).$$

We now give some elementary properties of homomorphisms.

Proposition 8.6 *Let G and H be groups and $\phi : G \to H$ be a homomorphism. Then for all x and y in G,*

$$\phi(xy^{-1}) = \phi(x)\phi(y)^{-1} \text{ and } \phi(y^{-1}x) = \phi(y)^{-1}\phi(x).$$

Proof We have

$$\begin{aligned} \phi(xy^{-1})\phi(y) &= \phi((xy^{-1})y) \quad \text{since } \phi \text{ is a homomorphism} \\ &= \phi(x). \end{aligned}$$

Now multiply both sides of this equation on the right by $\phi(y)^{-1}$ to obtain the result. The proof of the other assertion is similar and is left as an exercise. □

Corollary 8.7 *With the notation of Proposition 8.6,*

(i) $\phi(1_G) = 1_H$; *and*
(ii) *for all* $g \in G, \phi(g^{-1}) = \phi(g)^{-1}$.

Proof (i) follows from the proposition by taking $x = y$.
(ii) follows taking $x = 1_G$ and $y = g$. □

Definition 8.8 *A homomorphism from G to H which is a bijection is an* isomorphism. *In that case, we say that G and H are* isomorphic, *and write $G \cong H$.*

Since isomorphic groups have the 'same' multiplication table, the distinction between isomorphic groups is not always made very sharply. More formally, we have the following result.

Proposition 8.9 *Isomorphism is an equivalence relation on the set of groups.*

Proof The map $id_G : G \to G$ defined by $id_G(g) = g$ for all g in G, is clearly a homomorphism and bijective, showing that $G \cong G$. If ϕ is an isomorphism from G to H, then ϕ^{-1} exists because ϕ is a bijection. To show that ϕ^{-1} is an isomorphism from H to G, recall that $\phi^{-1}(y) = x$ whenever $\phi(x) = y$. Thus if also $\phi(x_1) = y_1$, so that $\phi(xx_1) = yy_1$, we have

$$\phi^{-1}(y)\phi^{-1}(y_1) = xx_1 = \phi^{-1}(yy_1),$$

showing that $\phi^{-1}$is a homomorphism. Finally if $\phi : G \to H$ and $\vartheta : H \to K$ are isomorphisms, it may easily be checked that $\vartheta \circ \phi$ is a homomorphism between G and K since

$$\begin{aligned}
\vartheta \circ \phi(xy) &= \vartheta(\phi(xy)) \\
&= \vartheta(\phi(x)\phi(y)) \quad \text{since } \phi \text{ is a homomorphism} \\
&= \vartheta(\phi(x))\vartheta(\phi(y)) \text{ since } \vartheta \text{ is a homomorphism} \\
&= (\vartheta \circ \phi(x))(\vartheta \circ \phi(y)).
\end{aligned}$$

The fact that $\vartheta \circ \phi$ is a bijection is easily checked directly, but was proved in detail in Corollary 2.16. □

Definition 8.10 *A bijective homomorphism ϕ from a group to itself is an* automorphism.

Proposition 8.11 *The set of automorphisms of a group G is itself a group, denoted by* Aut(G)*, under composition of maps.*

Proof We check the group axioms. If $\vartheta, \phi \in \text{Aut}(G)$, as we saw in the proof of Proposition 8.9, the composite $\vartheta \circ \phi$ is a homomorphism and is bijective because ϕ and ϑ are. The operation of composition is associative by Proposition 2.10. The identity map is bijective. If ϕ is in Aut(G), then ϕ^{-1} exists since ϕ is bijective, and ϕ^{-1} is a homomorphism as in the proof of Proposition 8.9. □

Definition 8.12 *Let $\phi : G \to H$ be a homomorphism. The* kernel $\ker \phi$ *is the set of elements of G which are mapped to the identity element of H:*

$$\ker \phi = \{g \in G : \phi(g) = 1_H\}.$$

The image $\text{im}\ \phi$ *is the set of elements of H which are images of elements of G under ϕ:*

$$\text{im}\ \phi = \{h \in H : h = \phi(g) \text{ for some } g \in G\}.$$

Theorem 8.13 (The Homomorphism Theorem) *Let ϕ be a homomorphism from G to H. Then*

(1) $\ker\phi$ *is a normal subgroup of* G;
(2) $\text{im }\phi$ *is a subgroup of* H; *and*
(3) $G/\ker\phi \cong \text{im }\phi$.

Proof (1) We first check that $\ker\phi$ is a subgroup of G. By Corollary 8.7(i), 1_G is in $\ker\phi$; if x and y are in $\ker\phi$, then

$$\phi(xy) = \phi(x)\phi(y) = 1_H 1_H = 1_H,$$

so $xy \in \ker\phi$; and if g is in $\ker\phi$, then $g^{-1} \in \ker\phi$ by Corollary 8.7(ii). To show that $\ker\phi$ is a normal subgroup, let $g \in G$ and $x \in \ker\phi$, then

$$\begin{aligned} \phi(g^{-1}xg) &= \phi(g)^{-1}\phi(x)\phi(g) \quad \text{using Proposition 8.6} \\ &= \phi(g)^{-1}1_H\phi(g) \quad \text{since } x \in \ker\phi \\ &= 1_H, \end{aligned}$$

so $g^{-1}xg$ is in $\ker\phi$, and $\ker\phi$ is a normal subgroup of G.

(2) Corollary 8.7(i) shows that 1_H is in $\text{im }\phi$. If y and y_1 are in $\text{im }\phi$ with $y = \phi(x)$ and $y_1 = \phi(x_1)$ for some x and x_1 in G, then

$$yy_1 = \phi(x)\phi(x_1) = \phi(xx_1),$$

so yy_1 is in $\text{im }\phi$. Also if y is in $\text{im }\phi$ with $y = \phi(x)$, then $y^{-1} = \phi(x^{-1})$, using Corollary 8.7(ii). Thus $\text{im }\phi$ is a subgroup of H.

(3) Write K for $\ker\phi$ for convenience. Define a map $\vartheta : G/K \to H$ by the rule

$$\vartheta(gK) = \phi(g) \quad \text{for all } g \in G.$$

To show that ϑ is an isomorphism, there are several things to check: that ϑ is well-defined, that ϑ is a homomorphism and finally that ϑ is a bijection from G/K to $\text{im }\phi$. The definition of ϑ appears to depend on the choice of coset representative, so to show that ϑ is well-defined, suppose that $xK = yK$, so that $x = yk$ for some $k \in K$, then

$$\vartheta(xK) = \phi(x) = \phi(yk) = \phi(y)\phi(k) = \phi(y)1_H = \phi(y) = \vartheta(yK),$$

so ϑ is well-defined. Next, for all x, y in G,

$$\vartheta(xK)\vartheta(yK) = \phi(x)\phi(y) = \phi(xy) = \vartheta(xyK),$$

so ϑ is a homomorphism. Since each element of $\text{im }\phi$ is $\phi(x) = \vartheta(xK)$ for some x in G, ϑ is a surjection from G/K to $\text{im }\phi$. It only remains to show that ϑ is injective: if $\vartheta(xK) = \vartheta(yK)$, we deduce that $\phi(x) = \phi(y)$, and so

$$1_H = \phi(y)^{-1}\phi(x) = \phi(y^{-1}x),$$

using Proposition 8.6. Thus $y^{-1}x \in K$ and so $xK = yK$. □

Example 8.14 Consider the homomorphism $\det : GL(n, \mathbf{R}) \to \mathbf{R}^\times$ taking each matrix to its determinant. The kernel of this map is the set $SL(n, \mathbf{R})$ of matrices of determinant 1. The map is surjective since there are invertible matrices of all possible non-zero determinants, so the homomorphism theorem shows that

$$GL(n, \mathbf{R})/SL(n, \mathbf{R}) \cong \mathbf{R}^\times.$$

Remark The homomorphism theorem tells us that the kernel of any homomorphism is a normal subgroup. The converse is true. If N is a normal subgroup of a group G, the map $\phi : G \to G/N$ defined by $\phi(g) = gN$, can easily be seen to be a homomorphism, known as the *natural homomorphism*. The kernel of this map is the normal subgroup N. There is therefore a correspondence between the set of normal subgroups of a group G and the kernels of homomorphisms from G. Similarly, there is a correspondence between the quotient groups of G and the isomorphism classes of the images of homomorphisms from G.

The next two results are important applications of the Homomorphism Theorem.

Theorem 8.15 (The First Isomorphism Theorem) *Let H be a subgroup of the group G and N be a normal subgroup of G. Then N is a normal subgroup of the group $\langle N, H\rangle = HN$ and $N \cap H$ is a normal subgroup of H. Furthermore*

$$\frac{H}{N \cap H} \cong \frac{HN}{N}.$$

Proof The fact that $\langle N, H\rangle = HN$ follows by Proposition 7.8, and the fact that N is a normal subgroup of G implies that N is a normal subgroup of HN. Now define a map $\phi : H \to HN/N$ by the rule $\phi(h) = hN$. (Note that H need not contain N so the coset hN is an element of the group HN/N rather than of H/N.) Then ϕ is a homomorphism since

$$\phi(xy) = xyN = (xN)(yN) = \phi(x)\phi(y).$$

The kernel of ϕ is given by

$$\ker\phi \quad = \quad \{h \in H : \phi(h) = 1_{HN/N}\}$$

$$\begin{aligned} &= \{h \in H : hN = N\} \\ &= \{h \in H : h \in N\} \\ &= H \cap N. \end{aligned}$$

Finally, ϕ is surjective since the element $hnN = hN$ of HN/N is $\phi(h)$. The result now follows by the Homomorphism Theorem. □

Theorem 8.16 (The Second Isomorphism Theorem) *Let H and N be normal subgroups of the group G with N contained in H. Then H/N is a normal subgroup of G/N, and*

$$(G/N)/(H/N) \cong G/H.$$

Proof Define a map $\phi : G/N \to G/H$ by $\phi(gN) = gH$. Since ϕ is defined on cosets, we should check that ϕ is well-defined. Suppose that $xN = yN$, so that $y^{-1}x \in N$. Then since $N \leq H$, we see that $y^{-1}x \in H$ and so $xH = yH$, so that $\phi(xN) = \phi(yN)$. Next, ϕ is a homomorphism since

$$\phi(xN)\phi(yN) = (xH)(yH) = xyH = \phi(xyN).$$

Also ϕ is clearly surjective and

$$\begin{aligned} \ker\phi &= \{gN \in G/N : \phi(gN) = 1_{G/H}\} \\ &= \{gN \in G/N : gH = H\} \\ &= \{gN \in G/N : g \in H\} \\ &= H/N, \end{aligned}$$

as required. The result now follows by Theorem 8.13. □

We conclude this section with another application of the Homomorphism Theorem. This will enable us to understand the automorphism group of a given group.

Proposition 8.17 *Let x be any element of a group G. Define a map $\phi_x : G \to G$ by $\phi_x(g) = xgx^{-1}$ for all $g \in G$.*

Then ϕ_x is an automorphism of G, known as the inner automorphism *of G given by x. Let* $\mathrm{Inn}(G)$ *denote the set of all inner automorphisms of G, and*

$$Z(G) = \{x \in G : gx = xg \text{ for all } g \in G\}.$$

Then Inn(G) *is a subgroup of* Aut(G); $Z(G)$ *is a normal subgroup of* G, *and* $G/Z(G) \cong$ Inn(G).

Proof To show ϕ_x is an automorphism requires three steps:

(a) ϕ_x is a homomorphism: for all $g, h \in G$

$$\phi_x(g)\phi_x(h) = (xgx^{-1})(xhx^{-1}) = x(gh)x^{-1} = \phi_x(gh);$$

(b) ϕ_x is injective:

$$\text{if } \phi_x(g) = \phi_x(h), \text{ then } xgx^{-1} = xhx^{-1}, \text{ and so } g = h;$$

(c) ϕ_x is surjective:

$$\text{given any } h \text{ in } G, \phi_x(x^{-1}hx) = x(x^{-1}hx)x^{-1} = h.$$

Now define a map $\vartheta : G \to \text{Aut}(G)$ by $\vartheta(x) = \phi_x$. We note that, for all g in G,

$$(\phi_x \circ \phi_y)(g) = \phi_x(\phi_y(g)) = \phi_x(ygy^{-1}) = xygy^{-1}x^{-1} = \phi_{xy}(g),$$

so $\phi_x \circ \phi_y = \phi_{xy}$. The fact that ϑ is a homomorphism then follows since

$$\vartheta(x)\vartheta(y) = \phi_x \circ \phi_y = \phi_{xy} = \vartheta(xy).$$

It is clear that the image of ϑ is Inn(G) and

$$\begin{aligned}
\ker \vartheta &= \{x \in G : \phi_x = id_G\} \\
&= \{x \in G : \phi_x(g) = id_G(g) \text{ for all } g \in G\} \\
&= \{x \in G : xgx^{-1} = g \text{ for all } g \in G\} \\
&= \{x \in G : xg = gx \text{ for all } g \in G\} \\
&= Z(G).
\end{aligned}$$

The Homomorphism Theorem now gives that $Z(G)$ is a normal subgroup of G, that Inn(G) is a subgroup of Aut(G), and also that the quotient group $G/Z(G)$ is isomorphic to Inn(G). □

Remark The subgroup $Z(G)$ occurring in Proposition 8.17 is known as the *centre* of the group G. It was introduced in Exercise 7.5.

Summary for Chapter 8

In this chapter, the idea of a *homomorphism* between groups G and H is defined. This is a map ϕ such that for all $x, y \in G, \phi(xy) = \phi(x)\phi(y)$. Special types of homomorphisms are: *isomorphisms* (when the map is bijective) and *automorphisms* (when $G = H$ and the map is bijective). The Homomorphism Theorem states that the kernel of a homomorphism $\phi : G \to H$ is a normal subgroup of G, that the image of ϕ is a subgroup of H, and that the quotient group $G/\ker\phi$ is isomorphic to $\text{im}\ \phi$. It follows from this that there is a one-to-one correspondence between kernels of homomorphisms and normal subgroups. It would therefore be possible to define the concept of normality by saying that a subgroup N of a group G is normal if and only if there is a homomorphism whose kernel is N. The chapter concludes with some applications of the Homomorphism Theorem. The first of these is to obtain the two Isomorphism Theorems. These will be required later. The second application is concerned with the structure of the automorphism group of a group.

Exercises 8

1. Let $\phi : G \to H$ be a homomorphism. Show that ϕ is injective if and only if $\ker\phi = \{1\}$.

2. Let G be the dihedral group $D(3)$. Define a map $\vartheta : G \to \{1, -1\}$ by $\vartheta(g) = 1$ if g is a rotation, and $\vartheta(g) = -1$ if g is a reflection. Prove that ϑ is a homomorphism, and calculate its kernel and image.

3. Suppose that H is an abelian group and let $\vartheta : G \to H$ be a homomorphism. Define a map $\phi : G \times G \to H$ by

$$\phi(g_1, g_2) = \vartheta(g_1)\vartheta(g_2)^{-1}.$$

Prove that ϕ is a homomorphism. List the elements in $\ker\ \phi$ when G is the dihedral group $D(3)$ and $\vartheta : G \to \{1, -1\}$ is the map of Question 2 above.

4. Let $\phi : G \to H$ be a homomorphism. Prove by induction that, for all positive integers k, and for all g in $G, \phi(g^k) = \phi(g)^k$. Deduce that if g has finite order k, then the order of $\phi(g)$ divides k, and that if also ϕ is injective, then the order of $\phi(g)$ is equal to k.

5. Determine the elements of Aut(G) when G is the cyclic group C_3 consisting of the three complex cube roots of unity, namely $1, \omega$ and ω^2, where $\omega = e^{2\pi i/3}$. Write down the multiplication table for Aut(G).

9

Permutations

In this chapter, we return to a situation which was discussed in Chapter 2. This is the idea of a *permutation of a set* X, that is, a bijection $\pi : X \to X$. We saw in Corollary 2.18, that the set $S(X)$ of permutations on X is a group under composition of maps. In this chapter, we shall be interested in the case where X has a finite number n, say, of elements. In this case, we shall usually regard X as the set $\{1, 2, \ldots, n\}$ and denote $S(X)$ as $S(n)$. It is common to write such a bijection $\pi : X \to X$ in *two-row* notation, in which the top row of a $2 \times n$ matrix contains the integers $1, 2, \ldots, n$ and the effect of π on the integer i is written under i:

$$\pi = \begin{pmatrix} 1 & 2 & 3 & \ldots & n \\ \pi(1) & \pi(2) & \pi(3) & \ldots & \pi(n) \end{pmatrix}.$$

Since the second row contains each element of X precisely once, there are n choices for the entry under 1, leaving $n-1$ choices for the entry under 2 and so on, giving $n!$ bijections altogether. Thus when $n = 3$, there are six permutations:

$$\begin{pmatrix} 1 & 2 & 3 \\ 1 & 2 & 3 \end{pmatrix}, \begin{pmatrix} 1 & 2 & 3 \\ 2 & 3 & 1 \end{pmatrix}, \begin{pmatrix} 1 & 2 & 3 \\ 3 & 1 & 2 \end{pmatrix},$$

$$\begin{pmatrix} 1 & 2 & 3 \\ 1 & 3 & 2 \end{pmatrix}, \begin{pmatrix} 1 & 2 & 3 \\ 3 & 2 & 1 \end{pmatrix}, \begin{pmatrix} 1 & 2 & 3 \\ 2 & 1 & 3 \end{pmatrix}.$$

A more compact notation is often used for elements of $S(n)$. It is called *cycle notation*, and it avoids repeating the same first row in each permutation. The theoretical basis for this notation is in the following result.

Proposition 9.1 *Let π be an element of $S(n)$ and i be any integer in $\{1, \ldots, n\}$. Let k be the smallest integer greater than 0 for which $\pi^k(i)$ is in the set $\{i, \pi(i), \ldots, \pi^{k-1}(i)\}$. Then $\pi^k(i) = i$.*

Proof Suppose that $\pi^k(i) = \pi^r(i)$ for some $r > 0$. Then, since π has an inverse, $\pi^{k-r}(i) = i$. This contradicts the definition of k, so $r = 0$. □

Definition 9.2 *A permutation ρ is a k*-cycle *if there exists a positive k and an integer i such that*

(1) *k is the smallest positive integer such that $\rho^k(i) = i$, and*
(2) *ρ fixes each j not in $\{i, \rho(i), \ldots, \rho^{k-1}(i)\}$.*

The k-cycle ρ is usually denoted $(i\ \rho(i)\ \ldots\ \rho^{k-1}(i))$.

Example 9.3 The five non-identity elements of $S(3)$ are all cycles, and may be written as

$$(1\ 2\ 3), (1\ 3\ 2), (2\ 3), (1\ 3) \text{ and } (1\ 2).$$

The identity permutation, fixing each of 1, 2 and 3, is usually denoted by 1. The permutation

$$\begin{pmatrix} 1 & 2 & 3 & 4 \\ 2 & 1 & 4 & 3 \end{pmatrix}$$

in $S(4)$ is not a cycle.

Definition 9.4 *Two permutations ρ, σ are* disjoint *if each number moved by ρ is fixed by σ, or equivalently, each number moved by σ is fixed by ρ.*

Proposition 9.5 *Every permutation may be written as a product of disjoint cycles.*

Proof To write a given element π of $S(n)$ as a product of disjoint cycles, start by producing a partition of $X = \{1, 2, \ldots, n\}$ as follows. First let $\mathrm{Fix}(\pi)$ denote the subset of X fixed by π, and form cycles of length 1, one for each $i \in \mathrm{Fix}(\pi)$. Next choose the smallest positive integer i, say, which is not in $\mathrm{Fix}(\pi)$. Let r be the smallest positive integer with $r > 1$ such that $\pi^r(i)$ is in the set $\{i, \pi(i), \ldots, \pi^{r-1}(i)\}$. Thus, by Proposition 9.1, $\pi^r(i) = i$, so we form the cycle

$$(i\ \pi(i)\ \ldots\ \pi^{k-1}(i)).$$

If there is any integer j, say, not fixed by π which is also not in the set $\{i, \pi(i), \ldots, \pi^{r-1}(i)\}$, let s be the smallest positive integer such that $\pi^s(j)$ belongs to the set $\{j, \pi(j), \ldots, \pi^{s-1}(j)\}$. Continue in this way to produce a partition of $\{1, \ldots, n\}$ into disjoint sets. [Alternatively, define a relation on $\{1, \ldots, n\}$ by iRj if and only if there is an integer k such that $\pi^k(i) = j$. It is easily checked that R is an equivalence relation and the classes are the required partition.] Continuing in this way for the sets in the partition, we may write π as a product of its disjoint cycles. □

Example 9.6 The cycle decomposition for the permutation

$$\pi = \begin{pmatrix} 1 & 2 & 3 & 4 & 5 & 6 & 7 & 8 & 9 \\ 1 & 4 & 6 & 2 & 8 & 9 & 7 & 5 & 3 \end{pmatrix}$$

is (1)(2 4)(3 6 9)(5 8). It is usual to omit the cycles of length 1, those integers fixed by π, and so π is abbreviated to (2 4)(3 6 9)(5 8).

It is just a convention that the smallest integer in each cycle is written first. From a notational point of view the cycles (3 6 9), (6 9 3) and (9 3 6) are equal, since each cycle sends 3 to 6, 6 to 9 and 9 to 3. However, all are different from the cycle (3 9 6).

Given a product of disjoint cycles, it is not clear which symmetric group the permutation belongs to. For example, we do not know which $S(n)$ the 2-cycle (1 2) belongs to since the fixed integers are not specified. However, it turns out to be an advantage to blur the distinction between elements of different symmetric groups in this way. We can multiply disjoint cycles without the need to interpret the permutations as elements of a particular symmetric group, just by composing the individual cycles. We illustrate this in the following example.

Example 9.7 Consider the product

$$\pi = (1\ 2\ 4)(3\ 5\ 7\ 9)(1\ 3\ 9)(2\ 3\ 4\ 5\ 6\ 8).$$

First consider the value $\pi(1)$. This is obtained by starting at the right-hand end of the decomposition and moving leftwards carrying 1 as the active symbol looking for the first 1. When we see this, the integer to the right of 1 (namely 3) becomes the active symbol. We move to the left looking for the first occurrence of 3, at this point 5 becomes the active symbol. Since no 5 occurs to the left of this point, the product takes 1 to 5, so that $\pi(1) = 5$. Now calculate $\pi(5)$ by starting at the right with 5 as the active symbol producing the change $5 \to 6$. Repeat again giving $6 \to 8$. Now repeat again with 8 as the active symbol. The first occurrence of 8 is at the right-hand end of a cycle, and so by construction of a cycle 8 is sent to 2, the integer at the left-hand end of its cycle, we therefore take 2 to be the active symbol and continue moving leftwards from the left end of the first cycle giving $8 \to 2 \to 4$. Continuing in this way, we can see that the given permutation is the 9-cycle (1 5 6 8 4 7 9 2 3).

A consequence of this rule is the following.

Proposition 9.8 *Let σ and ρ be disjoint permutations. Then $\sigma\rho = \rho\sigma$, and for all positive integers k, $(\sigma\rho)^k = \sigma^k\rho^k$. Let π be a product of disjoint* *cycles of lengths* $k_1, k_2, \ldots, k_r$, *then the order of* π *is the lowest common multiple of the integers* $k_1, k_2, \ldots, k_r$.

Proof We first point out that disjoint permutations commute. If ρ fixes i then $\sigma\rho(i) = \sigma(i)$, whereas $\rho\sigma(i) = \rho(\sigma(i))$. Since ρ and σ are disjoint, ρ must fix $\sigma(i)$ and so $\sigma\rho(i) = \rho\sigma(i)$. On the other hand, if ρ does not fix i then the fact that ρ and σ are disjoint means that σ must fix i, and it again follows that $\sigma\rho(i) = \rho\sigma(i)$.

Since σ and ρ commute, an inductive proof shows that for all positive integers k, $(\sigma\rho)^k = \sigma^k\rho^k$. This is done in two stages. First show by induction on k that if $\sigma\rho = \rho\sigma$ then $\sigma\rho^k = \rho^k\sigma$: when $k = 1$ this follows from the hypothesis; if we suppose that $\sigma\rho^k = \rho^k\sigma$, then

$$\begin{aligned}\sigma\rho^{k+1} &= \sigma\rho^k\rho = \rho^k\sigma\rho \\ &= \rho^k\rho\sigma = \rho^{k+1}\sigma.\end{aligned}$$

It then follows, also by induction on k, that $(\sigma\rho)^k = \sigma^k\rho^k$: this is clear when $k = 1$; if $(\sigma\rho)^k = \sigma^k\rho^k$, then

$$\begin{aligned}(\sigma\rho)^{k+1} &= (\sigma\rho)^k\sigma\rho = \sigma^k\rho^k\sigma\rho \\ &= \sigma^k\sigma\rho^k\rho = \sigma^{k+1}\rho^{k+1},\end{aligned}$$

as required.

Note that the order of a k-cycle is clearly k. Now suppose that π is a product of disjoint cycles

$$\rho_1, \rho_2, \ldots, \rho_r$$

of lengths $k_1, k_2, \ldots, k_r$, respectively. Let t be the lowest common multiple of $k_1, k_2, \ldots, k_r$. It then follows from the above that

$$\pi^t = \rho_1^t\rho_2^t \cdots \rho_r^t,$$

and so $\pi^t = 1$. The order of π therefore divides t. However if $\pi^s = 1$, then because the cycles are disjoint, each $\rho_i^s = 1$ and so s is divisible by each k_i, and so s is divisible by t. Thus the order of π is t. □

Example 9.9 The order of the permutation (1 2 3 4)(5 6 7)(8 9) is the lowest common multiple of 4, 3 and 2, so it is 12. However, the permutation (1 2 3 4)(2 6 7)(3 9) is not a product of disjoint cycles (and so need not have order 12). In fact,

$$(1\ 2\ 3\ 4)(2\ 6\ 7)(3\ 9) = (1\ 2\ 6\ 7\ 3\ 9\ 4),$$

so the permutation has order 7.

We have shown that every permutation can be written as a product of disjoint cycles. This decomposition is essentially unique: up to order of factors and ways of writing the same cycle. It is also possible to write each cycle as a product of *transpositions*, that is, cycles of length two.

Proposition 9.10 *Every k-cycle in $S(n)$ can be written as a product of $(k-1)$ transpositions.*

Proof The cycle $\pi = (1\ 2 \ldots\ k)$ has the factorisation

$$\pi = (1\ k) \ldots (1\ 3)(1\ 2),$$

so the general k-cycle $(i_1\ i_2\ \ldots\ i_k)$ factorises as

$$(i_1\ i_k) \ldots (i_1\ i_3)(i_1\ i_2),$$

giving the required factorisation of π. □

The factorisation of a cycle into transpositions is not unique. It is not even true that the number of transpositions in any factorisation of a given cycle is always the same, for example $(1\ 3) = (2\ 3)(1\ 2)(2\ 3)$. However, our next objective will be to show that the numbers of transpositions in any two decompositions of a given permutation are either both even or both odd.

Definition 9.11 *Let n be any positive integer, and let $f(x_1, x_2, \ldots, x_n)$ be a polynomial in n variables $x_1, x_2, \ldots, x_n$. Given an element π in $S(n)$, we define $\pi \cdot f$ to be the polynomial obtained by applying the permutation π to the subscripts on the variables.*

Example 9.12 If $\pi = (1\ 2\ 3)$ in $S(3)$ and

$$f(x_1, x_2, x_3) = x_1^2 + 2x_1x_2 - 4x_1x_2x_3^3,$$

then

$$\pi \cdot f = x_2^2 + 2x_2x_3 - 4x_2x_3x_1^3.$$

The following properties are easily checked.

Proposition 9.13 *Let π, ρ be elements of $S(n)$ and let f be any polynomial in variables $x_1, x_2, \ldots, x_n$. Then*

(a) $id \cdot f = f$;
(b) $\pi\rho \cdot f = \pi \cdot (\rho \cdot f)$; *and*
(c) *for all real numbers λ, $\pi \cdot (\lambda f) = \lambda(\pi \cdot f)$.* □

Remark This is an example of a group acting on a set (the set of polynomials in $x_1, x_2, \ldots, x_n$). We shall return to this idea in Chapter 10. We now apply this to a particular polynomial.

Definition 9.14 *For any positive integer n, let Δ_n be the polynomial in n variables $x_1, x_2, \ldots, x_n$ defined by*

$$\Delta_n(x_1, x_2, \ldots, x_n)$$

$$= (x_1 - x_2)(x_1 - x_3)\ldots(x_1 - x_n)(x_2 - x_3)\ldots(x_2 - x_n)\ldots(x_{n-1} - x_n)$$

$$= \prod_{1 \le i < j \le n} (x_i - x_j).$$

For any permutation π in $S(n), \pi \cdot \Delta_n$ is the polynomial

$$(x_{\pi(1)} - x_{\pi(2)})(x_{\pi(1)} - x_{\pi(3)})\ldots(x_{\pi(1)} - x_{\pi(n)})\ldots(x_{\pi(n-1)} - x_{\pi(n)})$$

$$= \prod_{1 \le i < j \le n} (x_{\pi(i)} - x_{\pi(j)}).$$

Remark In the above definition, each factor on the right-hand side of $\pi \cdot \Delta_n$ will occur, possibly with a sign change, in the original polynomial Δ_n. More precisely, the expression $(x_{\pi(i)} - x_{\pi(j)})$ will occur in Δ_n if $\pi(i) < \pi(j)$, whereas if $\pi(j) < \pi(i)$ then $-(x_{\pi(i)} - x_{\pi(j)})$ is in Δ_n.

Definition 9.15 *For any $\pi \in S(n)$, Δ_n is either equal to $\pi \cdot \Delta_n$, in which case we say that π is* even, *or $\Delta_n = -\pi \cdot \Delta_n$, in which case, we say that π is* odd. *We write* $\operatorname{sgn}(\pi) = 1$ *if π is even and* $\operatorname{sgn}(\pi) = -1$ *if π is odd, so that* $\pi \cdot \Delta_n = \operatorname{sgn}(\pi)\Delta_n$. *We refer to* $\operatorname{sgn}(\pi)$ *as the* sign *of π.*

Proposition 9.16 *The map* sgn: $S(n) \to C_2$ *is a homomorphism.*

Proof We must show that $\operatorname{sgn}(\pi\rho) = \operatorname{sgn}(\pi) \operatorname{sgn}(\rho)$. The proof of this is as follows:

$$\begin{aligned}
\operatorname{sgn}(\pi\rho)\Delta_n &= \pi\rho \cdot (\Delta_n) && \text{by definition} \\
&= \pi \cdot (\rho \cdot \Delta_n) && \text{by Proposition 9.13(b)} \\
&= \pi \cdot (\operatorname{sgn}(\rho)\Delta_n) && \text{by definition} \\
&= \operatorname{sgn}(\rho)(\pi \cdot \Delta_n) && \text{by Proposition 9.13(c)}
\end{aligned}$$

$$= \quad \text{sgn}(\rho)\text{sgn}(\pi)\Delta_n \quad \text{by definition.}$$

Thus $\text{sgn}(\pi\rho) = \text{sgn}(\rho)\ \text{sgn}(\pi) = \text{sgn}(\pi)\ \text{sgn}(\rho)$, as required. □

Corollary 9.17 *For any element π in $S(n)$, $\text{sgn}(\pi^{-1}) = \text{sgn}(\pi)$. For any π, ρ in $S(n)$,*

$$\text{sgn}(\pi\rho\pi^{-1}) = \text{sgn}(\rho).$$

Proof It follows immediately from the definition that $\text{sgn}(id) = 1$. Since $id = \pi\pi^{-1}$, the proposition shows that $\text{sgn}(\pi)\ \text{sgn}(\pi^{-1}) = 1$, from which it follows that $\text{sgn}(\pi^{-1}) = \text{sgn}(\pi)$. Since

$$\text{sgn}(\rho)\text{sgn}(\pi) = \text{sgn}(\pi)\text{sgn}(\rho)$$

it follows that

$$\text{sgn}(\pi\rho\pi^{-1}) = \text{sgn}(\pi)\text{sgn}(\rho)\text{sgn}(\pi) = \text{sgn}(\rho),$$

as claimed. □

Corollary 9.18 *A transposition is odd. A k-cycle is even if and only if k is an odd integer.*

Proof We show that a transposition is odd in a sequence of steps. First, the fact that the transposition $(1\ 2)$ is odd follows from the definition of sgn using Δ_n. Interchanging 1 and 2 only produces one sign change in $(1\ 2) \cdot \Delta_n$, that occurring in the factor $(x_1 - x_2)$. Next a transposition of the form $(1\ k)$ may be written as

$$(1\ k) = (2\ k)(1\ 2)(2\ k)^{-1},$$

and so is odd by Corollary 9.17. Finally,

$$(\ell\ k) = (1\ \ell)(1\ k)(1\ \ell)^{-1}$$

so another application of Corollary 9.17 shows that any transposition is odd. By Proposition 9.10, a k-cycle is a product of $k-1$ odd permutations and so is even if and only if k is an odd integer. □

Definition 9.19 *The kernel of the map sgn is the* alternating group $A(n)$.

Remark By the Homomorphism Theorem and the fact that sgn is onto C_2, the alternating group consists of the set of even permutations in $S(n)$. It also follows that $A(n)$ is a normal subgroup of index 2 when $n \geq 2$. In particular, $A(3)$ consists of the three permutations $\{id, (1\ 2\ 3), (1\ 3\ 2)\}$. We shall see later that, for $n \geq 5$, the only normal subgroups of $A(n)$ are $\{1\}$ and $A(n)$ itself.

The following result will be very useful in future calculations when we wish to determine the list of elements in $S(n)$ conjugate to a given permutation.

Proposition 9.20 *Let π and ρ be permutations in $S(n)$. The cycle decomposition of the element $\pi\rho\pi^{-1}$ is obtained from that of ρ by replacing each integer i in the cycle decomposition of ρ with the integer $\pi(i)$.*

Proof Consider the effect that $\pi\rho\pi^{-1}$ has on the integer $\pi(i)$:

$$\pi\rho\pi^{-1}(\pi(i)) = \pi(\rho(i)).$$

In other words, $\pi\rho\pi^{-1}$ maps $\pi(i)$ to $\pi(\rho(i))$: in the cycle decomposition of $\pi\rho\pi^{-1}$, $\pi(i)$ stands to the left of $\pi(\rho(i))$, whereas in the cycle decomposition for ρ, i stands to the left of $\rho(i)$. □

Remark This proposition shows that if two permutations are conjugate, their decompositions into disjoint cycles must have the same numbers of cycles of each length. We express this by saying that conjugate permutations have the same *cycle type*. The proposition can also be used in the reverse direction. Given two permutations σ, ϕ in $S(n)$ with the same cycle type, there is an element $\pi \in S(n)$ such that $\phi = \pi\sigma\pi^{-1}$. For example, conjugating the permutation $(1\ 2\ 3\ 4\ 5)$ by $(1\ 2)(3\ 4)$, the proposition shows that the result is $(2\ 1\ 4\ 3\ 5) = (1\ 4\ 3\ 5\ 2)$. Similarly the permutation $(1\ 2)(3\ 4)$ is conjugate to $(1\ 3)(2\ 5)$ via

$$\begin{pmatrix} 1 & 2 & 3 & 4 & 5 \\ 1 & 3 & 2 & 5 & 4 \end{pmatrix}.$$

Our next result in this chapter is an application of Proposition 9.20.

Proposition 9.21 *For any integer n, the transpositions $a_k = (k\ k+1)$ for $(1 \leq k \leq n-1)$ are a set of generators for $S(n)$ satisfying the relations*

$$\begin{aligned} a_k^2 &= 1, \quad (1 \leq k \leq n-1); \\ (a_k a_{k+1})^3 &= 1, \quad (1 \leq k \leq n-2); \\ (a_i a_j)^2 &= 1, \quad (1 \leq i, j \leq n-1 \text{ and } |i-j| > 1). \end{aligned}$$

Proof We first show that each transposition $(i\ j)$ with $i < j$ is in the subgroup $\langle a_1, a_2, \ldots, a_{n-1}\rangle$. Proposition 9.20 may be used to see that conjugating a_i by a_{i+1} gives $(i\ i+2)$, and in fact conjugating a_i by the product $a_{j-1}a_{j-2}\ldots a_{i+1}$ will give $(i\ j)$. Since every cycle is a product of transpositions by Proposition 9.10, and every permutation is a product of cycles by Proposition 9.5, we see that

$$\langle a_1, a_2, \ldots, a_{n-1}\rangle = S(n).$$

The given relations for $a_1, a_2, \ldots, a_{n-1}$ are easily checked since a_i and a_j are disjoint if $|i - j| > 1$ and $a_i a_{i+1}$ is the 3-cycle $(i\ i+1\ i+2)$. □

We conclude this chapter with a discussion of the connections between 'abstract' groups and permutation groups. The proof to be given is not just for finite groups.

Proposition 9.22 *Let G be a group and H be any subgroup of G. Let $\mathcal{S}$ be the set of distinct left cosets of H in G. For any g in G the map $\vartheta_g : \mathcal{S} \to \mathcal{S}$ defined by $\vartheta_g(xH) = gxH$ is a permutation of $\mathcal{S}$. The map ϑ defined by $\vartheta(g) = \vartheta_g$ is a homomorphism from G into the symmetric group on $\mathcal{S}$, and the kernel of this map is the subgroup*

$$\bigcap_{x\in G} xHx^{-1}.$$

Proof We must first show that ϑ_g is well-defined: if $xH = yH$, so that $y^{-1}x$ is in H, then $(gy)^{-1}gx = y^{-1}x \in H$, and so $\vartheta_g(yH) = \vartheta_g(xH)$. Reversing the order of steps in this argument shows that ϑ_g is injective. Finally, ϑ_g is surjective since for any left coset $xH, \vartheta_g(g^{-1}xH) = xH$.

Now to show that ϑ is a homomorphism, consider

$$\vartheta_{uv}(xH) = uvxH = \vartheta_u(vxH) = \vartheta_u(\vartheta_v(xH)).$$

This shows that $\vartheta_{uv} = \vartheta_u\vartheta_v$, so that $\vartheta(uv) = \vartheta(u)\vartheta(v)$, as required.

The final step in the proof is the calculation of the kernel of ϑ.

$$\begin{aligned}
\ker(\vartheta) &= \{g \in G : \vartheta_g = id\}\\
&= \{g \in G : \vartheta_g(xH) = xH \text{ for all } x \in G\}\\
&= \{g \in G : gxH = xH \text{ for all } x \in G\}\\
&= \{g \in G : x^{-1}gxH = H \text{ for all } x \in G\}\\
&= \{g \in G : x^{-1}gx \in H \text{ for all } x \in G\}
\end{aligned}$$

$$= \{g \in G : g \in xHx^{-1} \text{ for all } x \in G\}$$

$$= \bigcap_{x \in G} xHx^{-1},$$

as required. □

Corollary 9.23 *Let H be a subgroup of a group G with finite index n. Then there exists a normal subgroup N of G contained in H with n dividing $|G : N|$ and $|G : N|$ dividing $n!$.*

Proof Apply Proposition 9.22 to the subgroup H and let N be the kernel of ϑ. Then N is a normal subgroup of G and H contains N so that H/N is a subgroup of G/N of index n, by Theorem 7.14. It follows that n divides $|G : N|$. Also, by Proposition 9.22, G/N is isomorphic to a subgroup of the symmetric group $S(n)$ and so $|G : N|$ divides $n!$, as required. □

Corollary 9.24 (Cayley's Theorem) *Every group is isomorphic to a subgroup of a symmetric group.*

Proof Apply the proposition to the case when H is the subgroup $\{1\}$, so that S is the set G and the kernel of the map ϑ is $\{1\}$. The result then follows by the Homomorphism Theorem. □

Our final result in this section uses Proposition 9.20 to generalise a fact which was discussed in Example 7.6 for $p = 2$.

Corollary 9.25 *Let G be a group and suppose that p is the smallest prime dividing $|G|$. If G has a subgroup H of index p, then H is a normal subgroup of G.*

Proof Apply Corollary 9.23 to the subgroup H to find a normal subgroup N with $|G : N|$ dividing $p!$. Since $|G : N|$ also divides $|G|$, it divides the greatest common divisor of $p!$ and $|G|$. Since p is the smallest prime dividing $|G|$, the greatest common divisor of $p!$ and $|G|$ is p, so $|G : N| = p = |G : H|$. Since N is contained in H, it follows that N is equal to H. □

This may be a good place to consider the group $S(4)$ in more detail. This group has 24 elements of which 6 are transpositions (2-cycles), 8 are 3-cycles, 6 are 4-cycles and the remaining 3 non-identity elements are products of two disjoint transpositions (for example $(1\ 2)(3\ 4)$). There are 12 even permutations in $S(4)$, these being the 3-cycles together with the identity element and the three products of disjoint transpositions. These 12 elements form a subgroup of index 2 which is therefore a normal subgroup by Example 7.6. It then follows by Proposition 5.19 that the factor group $S(4)/A(4)$ is cyclic of order 2. It may be checked easily that the identity element together with the three products of disjoint transpositions form a subgroup V. It then follows by Proposition 9.20 that each conjugate of a non-identity element of V is again an element of v, so that V is, in fact, a normal subgroup of G. The factor group $S(4)/V$ is non-abelian of order 6 and is isomorphic to $S(3)$ using the results of Chapter 5.

Summary for Chapter 9

In this chapter, we return to the topic of permutations which was briefly discussed in Chapter 2. There are two standard notations for permutations, the two-row notation and cycle notation. When a permutation is written as a product of disjoint cycles, it is easy to calculate its order: this is the lowest common multiple of the lengths of the cycles in the disjoint decomposition. We also introduce the idea of the sign of a permutation, and show that the sign map, sgn, is a group homomorphism. The kernel of this map is the set of even permutations, and this is known as the alternating group $A(n)$. One result which is used frequently later is that the conjugate permutation $\rho\pi\rho^{-1}$ is obtained by replacing each entry i in the cycle decomposition of π by $\rho(i)$.

As we have seen, when Galois and Cauchy introduced the concept of a group, the idea was restricted to groups of permutations. In the 1850s Cayley considered general sets with an associative operation with identity. This was a forerunner of the theory of 'abstract groups'. The result known as Cayley's Theorem, Corollary 9.24, shows that every abstract group is isomorphic to a group of permutations. Many of the basic results on permutations were established by Cauchy during the 1840s.

Exercises 9

1. Let π and ρ be the permutations

$$\pi = \begin{pmatrix} 1 & 2 & 3 & 4 & 5 & 6 & 7 & 8 \\ 3 & 4 & 5 & 6 & 7 & 2 & 1 & 8 \end{pmatrix} \text{ and } \rho = \begin{pmatrix} 1 & 2 & 3 & 4 & 5 & 6 & 7 & 8 \\ 8 & 7 & 6 & 5 & 4 & 3 & 2 & 1 \end{pmatrix}.$$

Write π and ρ in cycle notation. Calculate $\pi\rho, \rho\pi$ and $\pi^2\rho$. Find the orders and signs of π, ρ and $\pi\rho$.

2. List all 24 elements of $S(4)$ in cycle notation. Show that the subset

$V = \{1, (1\ 2)(3\ 4), (1\ 3)(2\ 4), (1\ 4)(2\ 3)\}$ is a subgroup of $S(4)$. Use the fact that any permutation π has the same cycle structure as $\rho\pi\rho^{-1}$ (for any ρ) to show that V is a normal subgroup of $S(4)$. Give names to the six left cosets of V, and write down the table for the quotient group $S(4)/V$.

3. List all the elements of $A(4)$ and determine their orders. Find the list of subgroups of $A(4)$. Deduce that the converse of Lagrange's Theorem is not true: there is a divisor d of the order of $A(4)$ for which $A(4)$ has no subgroup with d elements.

10

The Orbit–Stabiliser Theorem

In this chapter, the concept of G-sets is introduced.

Definition 10.1 *Given a group G, a G-set is a set X together with a rule assigning, to each element g of G and each element x of X, an element $g \cdot x$ of X satisfying the requirements:*

(G-set 1) *for all x in X, $1_G \cdot x = x$;*

(G-set 2) *for all g_1 and g_2 in G and all x in X,*

$$(g_1 g_2) \cdot x = g_1 \cdot (g_2 \cdot x).$$

Example 10.2 An example of G-sets which we have already seen occurs when G is a subgroup of the group of permutations of the set X. For example, if $G = S(n)$ and $X = \{1, 2, \ldots, n\}$, the identity permutation takes each element of X to itself and condition (G-set 2) is a consequence of the multiplication rule in $S(n)$.

Example 10.3 We saw another example of an $S(n)$-set in Chapter 9. This was the set of all real polynomials in n variables $x_1, x_2, \ldots, x_n$, the action of a permutation π on such a polynomial being to permute the variables according to π. The conditions (G-set 1) and (G-set 2) follow immediately from the definitions as discussed in Proposition 9.13.

Example 10.4 Given a group G, we can make G into a G-set (that is, take X to be G) by defining $g \cdot x$ to be the group product gx.

Example 10.5 Another way to make G into a G-set is to use conjugation to define $g \cdot x$ to be gxg^{-1}. Clearly $1 \cdot x$ is x and

$$\begin{aligned}(g_1 g_2) \cdot x &= (g_1 g_2) x (g_1 g_2)^{-1} \\ &= g_1 g_2 x g_2^{-1} g_1^{-1}\end{aligned}$$

$$= g_1 \cdot (g_2 x g_2^{-1})$$

$$= g_1 \cdot (g_2 \cdot x),$$

which checks (G-set 2).

Example 10.6 As our next example, let X be the set of all subsets of the group G. For any subset H of G and any element x of G define $x \cdot H$ to be xH. It can easily be checked that the G-set requirements are satisfied.

Example 10.7 There is also a conjugation action of G on the set of all subgroups of G. In this case $x \cdot H$ is defined to be xHx^{-1}. The check that this satisfies (G-set 1) and (G-set 2) is similar to that in Example 10.5

Definition 10.8 *Given a G-set X, and an element x of X, the* stabiliser *G_x, is the set of group elements which fix x:*

$$G_x = \{g \in G : g \cdot x = x\}.$$

Proposition 10.9 *For any G-set X and any x in X, the stabiliser G_x is a subgroup of G.*

Proof Condition (G-set 1) ensures that the identity element of G is in G_x. If g_1 and g_2 are in G_x, so that $g_1 \cdot x = x = g_2 \cdot x$, then condition ($G$-set 2) ensures that

$$(g_1 g_2) \cdot x = g_1 \cdot (g_2 \cdot x) = g_1 \cdot x = x,$$

so that $g_1 g_2 \in G_x$. Also if g is in $G_x, g \cdot x = x$, and so

$$g^{-1} \cdot x = g^{-1} \cdot (g \cdot x) = (g^{-1} g) \cdot x = 1_G \cdot x = x,$$

showing that $g^{-1} \in G_x$. □

Example 10.10 We discuss stabilisers in each of the above six examples. In $S(n)$, the stabiliser of an integer i is the set of permutations which fix i. Thus if i is taken to be n, the stabiliser is the subgroup $S(n-1)$ of $S(n)$. In Example 10.3, the stabiliser of a polynomial is the set of permutations which fix the given polynomial. In Example 10.4, $G_x = \{1_G\}$, since the equation $gx = x$ implies that $g = 1$. In Example 10.5, the stabiliser is more commonly known as the *centraliser* $C_G(x)$, so that

$$C_G(x) = \{g \in G : gxg^{-1} = x\} = \{g \in G : gx = xg\}.$$

Since the powers of x commute with x, $\langle x \rangle$ is contained in $C_G(x)$. In Example 10.6, the stabiliser of H is the set $\{g \in G : gH = H\}$. When H is a subgroup, this is the set H. Finally, in Example 10.7, the stabiliser of H is known as the *normaliser* $N_G(H)$, the set $\{g \in G : gHg^{-1} = H\}$. Note that $N_G(H)$ always contains H and that H is a normal subgroup if and only if $N_G(H) = G$.

Example 10.11 Let H be the subgroup $\langle b \rangle$ in

$$D(3) = \langle a, b : b^2 = 1 = a^3 \text{ and } ab = ba^{-1} \rangle.$$

Then $N_G(H)$ contains H, and is a subgroup of G, and so has order 2 or 6. However H is not a normal subgroup of G, so $N_G(H) \neq G$ and so $N_G(H) = H$.

Definition 10.12 *Given a G-set X, define a relation R on X by xRy if and only if there exists an element g in G with $y = g \cdot x$.*

Proposition 10.13 *The relation R is an equivalence relation.*

Proof Condition (G-set 1) ensures that xRx. If xRy, so that $y = g \cdot x$ for some $g \in G$, then $g^{-1} \cdot y = g^{-1} \cdot (g \cdot x) = x$, using ($G$-set 2), so yRx. Finally if xRy and yRz, with $y = g \cdot x$ and $z = h \cdot y$, then

$$z = h \cdot (g \cdot x) = hg \cdot x,$$

and so xRz. The equivalence class of x is the set of all elements of X of the form $\{g \cdot x : g \in G\}$. □

Definition 10.14 *The equivalence class of $x \in X$ under R is known as* the orbit *of x, and is given by* $\text{orb}(x) = \{g \cdot x : g \in G\}$.

Example 10.15 We look at orbits in some of our standard examples. In Example 10.4, the orbit of x is the whole group G by Proposition 3.3. In Example 10.5, the orbit of x is known as the *conjugacy class of x*; it is the set of all group elements of the form $\{gxg^{-1}\}$ as g varies over G. In Example 10.6, the orbit of a subgroup H is the set of left cosets of H in G and under the conjugation action, the orbit is the set of *subgroups conjugate to* $H, \{gHg^{-1} : g \in G\}$.

Theorem 10.16 (The Orbit–Stabiliser Theorem) *Let G be a group and X be a G-set. For each x in X,*

$$|\text{orb}(x)| = |G : G_x|.$$

Proof Given any $x \in X$, define a map ϑ from the orbit of x to the set of left cosets of G_x in G, by $\vartheta(g \cdot x) = gG_x$. (Note this is not a homomorphism: orb(x) is not a group!) We first show that ϑ is well-defined in that its value does not depend on the choice of g to obtain an element in the orbit of x. Suppose, therefore, that $g \cdot x = h \cdot x$. It then follows, acting on both sides on the left by h^{-1}, that $h^{-1}g \cdot x = x$. Thus $h^{-1}g \in G_x$ so that, by Corollary 5.5, $gG_x = hG_x$, as required.

We next show that ϑ is injective. If $\vartheta(g_1 \cdot x) = \vartheta(g_2 \cdot x)$, then $g_1G_x = g_2G_x$, so by Corollary 5.5, $g_2^{-1}g_1 \in G_x$. Thus, by definition of G_x,

$$x = g_2^{-1}g_1 \cdot x.$$

Applying g_2 to both sides gives that

$$\begin{aligned} g_2 \cdot x &= g_2 \cdot (g_2^{-1}g_1 \cdot x) \\ &= (g_2g_2^{-1}g_1) \cdot x \quad \text{by } (G\text{-set } 2) \\ &= g_1 \cdot x, \end{aligned}$$

and so ϑ is injective. It is clear that ϑ is surjective since the left coset gG_x is $\vartheta(g \cdot x)$ by definition. □

Remark When applied to Example 10.6 above, the Orbit–Stabiliser Theorem gives that for any subgroup H, the number of distinct left cosets of H in G is equal to $|G : H|$. Thus Lagrange's Theorem is a special case of Theorem 10.16.

Example 10.17 We calculate the conjugacy classes in the group

$$D(3) = \langle a, b : b^2 = 1 = a^3 \text{ and } ab = ba^{-1} \rangle.$$

This means that, for each x in $D(3)$, we will determine the complete list of distinct elements of the form gxg^{-1} as g varies over $D(3)$. The importance of the Orbit–Stabiliser Theorem for this calculation is that it tells how many elements to expect in the conjugacy class of x, namely $|G|/|C_G(x)|$.

In any group, the identity element always forms a conjugacy class since, for all $g \in G$, $g1g^{-1} = 1$. To calculate the conjugacy class containing a, note that since $b^2 = 1$, the conjugate $b^{-1}ab = bab^{-1} = a^{-1}$, so $a^{-1} = a^2$ is in the conjugacy class of a. Since $C_G(a)$ contains $\langle a \rangle$ and is a subgroup of G, its order is divisible by 3 and divides 6. However, $C_G(a)$ is not G since there is an element g in G with $ag \neq ga$ (take $g = b$). It follows that $C_G(a)$ has three elements and so by Theorem 10.16, a has two conjugates. Hence $\{a, a^2\}$ is a conjugacy class in G. We next calculate the conjugacy class containing b. An argument similar to the above shows that $C_G(b)$ is $\langle b \rangle$ and so has two elements. This means that b has three conjugates. Since conjugacy classes partition G, the only possible elements in the class of b are b, ba and ba^2. Thus the conjugacy classes of G are

$$\{1\}, \{a, a^2\} \text{ and } \{b, ba, ba^2\}.$$

The following fact is often useful when calculating conjugacy classes.

Proposition 10.18 *Conjugate elements have the same order.*

Proof We first show, by induction on n, that $(gxg^{-1})^n = gx^n g^{-1}$. This is clear when $n = 1$. Suppose it is true for $n = k$, then

$$\begin{aligned}(gxg^{-1})^{k+1} &= (gxg^{-1})^k gxg^{-1} \\ &= gx^k g^{-1} gxg^{-1} \\ &= gx^{k+1} g^{-1},\end{aligned}$$

as required. If now x has order n, then we see that

$$(gxg^{-1})^n = gx^n g^{-1} = 1,$$

so that the order of gxg^{-1} divides n. However, if the order of gxg^{-1} is k, it follows from the fact that $x = g^{-1}(gxg^{-1})g$ that the order of x divides k. We deduce that x and gxg^{-1} have the same order. □

Remark 1 An alternative proof of Proposition 10.18 follows from Exercise 8.4, by taking ϕ to be the inner automorphism induced by conjugation by g.

Remark 2 The converse of Proposition 10.18 is not true. In the dihedral group $D(4)$, rotation through 180° has order 2, as does any reflection. But this rotation is not conjugate to any reflection because each conjugate is equal to the rotation itself.

We next give a simple fact which is often useful in relating conjugacy classes of a group to those of a subgroup.

Proposition 10.19 *Let H be a subgroup of the group G. Then, for any x in G,*

$$C_H(x) = C_G(x) \cap H.$$

Proof It is clear that g is in $C_H(x)$ if and only if $g \in C_G(x)$ and g is in H. □

We next use these ideas to classify groups with p^2 elements, where p is any prime integer.

Recall that for any group G, the *centre* of G, $Z(G)$, is the set of all those elements which commute with every element of G:

$$Z(G) = \{z \in G : zx = xz \text{ for all } x \text{ in } G\}.$$

It was shown in Proposition 8.17 that $Z(G)$ is a normal abelian subgroup of G.

Proposition 10.20 *Let p be a prime integer and let G be a finite group with p^n elements. Then $Z(G)$ contains more than one element.*

Proof Let $\mathcal{C}_1, \mathcal{C}_2, \ldots, \mathcal{C}_r$ be the conjugacy classes of G. Since these are the equivalence classes under the relation of conjugacy, they partition G so that

$$|G| = |\mathcal{C}_1| + |\mathcal{C}_2| + \cdots + |\mathcal{C}_r|. \tag{10.1}$$

We number the classes so that $\mathcal{C}_1 = \{1\}$. By the Orbit–Stabiliser Theorem, if x_i is an element in $\mathcal{C}_i$, $|\mathcal{C}_i| = |G|/|C_G(x_i)|$. Thus $|\mathcal{C}_i|$ is a power of p, which depends on i and may possibly be 1 ($= p^0$) if $C_G(x_i) = G$. It is not possible for p to divide all the integers $|\mathcal{C}_i|$ for $2 \leq i \leq r$, because then the right-hand side of (10.1) would be $\equiv 1 \bmod p$, whereas the left-hand side is divisible by p. It therefore follows that there exists a value of i (other than $i = 1$) for which $C_G(x_i) = G$. Then $gx_i = x_i g$ for all x in G, so $x_i \in Z(G)$, showing that $Z(G)$ has more than one element. □

Proposition 10.21 *Let G be a group such that $G/Z(G)$ is cyclic. Then G is abelian, so that $G = Z(G)$.*

Proof Suppose that $G/Z(G)$ is cyclic generated by $xZ(G)$, so that every element of G lies in one of the cosets $Z(G), xZ(G), x^2Z(G), \ldots$ Take two arbitrary elements of G, $x^i z$ and $x^j w$ for some i, j and some z, w in $Z(G)$. Then

$$(x^i z)(x^j w) = x^i (z x^j) w = x^i (x^j z) w = x^{i+j} z w = x^{j+i} w z = (x^j w)(x^i z).$$

Thus G is abelian and so $G = Z(G)$. □

Corollary 10.22 *Let p be a prime integer. Any group with p^2 elements is abelian.*

Proof By Proposition 10.20, $Z(G)$ has more than one element, and so by Lagrange's Theorem, $Z(G)$ has p or p^2 elements. If $|Z(G)| = p^2$, then $Z(G) = G$ and so G is abelian. If $|Z(G)| = p$, then $G/Z(G)$ would have p elements and so would be cyclic by Proposition 5.19. It would then follow by Proposition 10.21 that $Z(G) = G$, so this case cannot arise. □

Corollary 10.23 *A group with p^2 elements is either cyclic or isomorphic to the direct product $C_p \times C_p$.*

Proof If G has p^2 elements, and G has an element of order p^2, then G is cyclic. Thus, if G is not cyclic, we may suppose that every non-identity element of G has order p. Let x be any such element, so that $\langle x \rangle$ has p elements. Now choose any element y of G not in $\langle x \rangle$. Since

$$\langle x \rangle \cap \langle y \rangle = \{1\},$$

it follows by Proposition 5.18 that the p^2 elements $\{x^i y^j : 0 \le i, j \le p-1\}$ are all distinct and so are all the elements of G. It may then be checked that the map

$$x^i y^j \mapsto (x^i, y^j).$$

is, in fact, an isomorphism between G and a direct product of the groups $\langle x \rangle$ and $\langle y \rangle$. □

We conclude this section with a generalisation of a fact we have already established.

Definition 10.24 *For any subgroup H of a group G, define the* centraliser *of H in G by*

$$C_G(H) = \{g \in G : ghg^{-1} = h \text{ for all } h \in H\}.$$

The following basic property of centralisers is easily checked.

Proposition 10.25 *Let H be a subgroup of a group G. Then for all elements g in G,*

$$C_H(x) = C_G(x) \cap H. \qquad \square$$

Proposition 10.26 *Let H be a subgroup of the group G. Then $C_G(H)$ is a normal subgroup of the group $N_G(H)$ and $N_G(H)/C_G(H)$ is isomorphic to a subgroup of* $\mathrm{Aut}(H)$.

Proof For each x in $N_G(H)$, define a map ϑ_x on H by $\vartheta_x(h) = xhx^{-1}$. Each ϑ_x is injective since if $xhx^{-1} = xkx^{-1}$ then $h = k$. It is also surjective since for any $k \in H$, $x^{-1}kx$ is in H because x^{-1} normalises H, and also $\vartheta_x(x^{-1}kx) = k$. To show that ϑ_x is an automorphism of H, therefore, it only remains to show that ϑ_x is a homomorphism. This follows since

$$\vartheta_x(hk) = x(hk)x^{-1} = (xhx^{-1})(xkx^{-1}) = \vartheta_x(h)\vartheta_x(k).$$

The map $x \mapsto \vartheta_x$ is a homomorphism since

$$\vartheta_x\vartheta_y(h) = \vartheta_x(yhy^{-1}) = xyhy^{-1}x^{-1} = (xy)h(xy)^{-1} = \vartheta_{xy}(h).$$

The kernel of this map is easily checked to be $C_G(H)$, and so the result follows from the Homomorphism Theorem, Theorem 8.13 and Proposition 10.25. $\square$

Summary for Chapter 10

In this chapter, we consider sets X equipped with a rule for combining an element g from G with an element x from X to produce an element $g \cdot x$ of X. There are two requirements on such a set in order that it is a G-set; the identity group element must fix each element of X, and for all $g, h \in G$ and all $x \in X$, we require that $g \cdot (h \cdot x) = gh \cdot x$. For any G-set X and any element x in X, we define the *stabiliser* G_x to be the set of group elements which fix x. This is shown (Proposition 10.9) to be a subgroup of G. There is also an equivalence relation xRy if and only if $y = g \cdot x$ for some $g \in G$. The equivalence classes under this relation are known as *orbits*. One of the most important examples of G-sets occurs when we take X to be G itself and define $g \cdot x$ to be gxg^{-1}. In this case the stabiliser G_x is also known as the *centraliser* $C_G(x)$ and is the set of all those group elements g such that $gx = xg$. The orbit of x in this case is known as the *conjugacy class* of x. The main result of this section shows that for any group G and any G-set X, for any $x \in X$, the number of elements in the orbit of x is equal to the index of G_x in G. Among the more important consequences of this result is the fact that every group whose order is a power of a prime has an element in its centre other than 1. This fact is used to classify groups with p^2 elements.

Exercises 10

1. Let G be the symmetric group $S(n)$ and V be a complex vector space with basis $\{v_1, v_2, \ldots, v_n\}$. For $\pi \in G$ and any $v = \lambda_1 v_1 + \cdots + \lambda_n v_n$ of V, define

$$\pi \cdot v = \lambda_1 v_{\pi(1)} + \cdots + \lambda_n v_{\pi(n)}.$$

Show that V is a G-set, and find both $\mathrm{orb}(v)$ and G_v when

(a) $n = 4$ and $v = v_1 + v_2 + v_3 + v_4$;
(b) $n = 4$ and $v = v_1 + v_3$.

2. Let X be a G-set and $x \in X$. Show that for any $g \in G$, the stabiliser of $g \cdot x$ is the subgroup gG_xg^{-1}.

3. Determine the list of conjugacy classes in the symmetric group $S(4)$ and also in the alternating group $A(4)$.

4. Let G be any group and g be an element of G. Prove directly that $C_G(g) = \{x \in G : gx = xg\}$ is a subgroup of G.

5. Let G be a finite group with precisely two conjugacy classes. Prove that G has two elements.

6. Let G be the group $S(3)$. Calculate $N_G(H)$ when H is

(a) the subgroup $\{1, (1\ 2\ 3), (1\ 3\ 2)\}$ and
(b) the subgroup $\{1, (1\ 2)\}$.

11

The Sylow Theorems

In this chapter, we shall prove the Sylow Theorems. These provide valuable insight into the structure of finite groups. In particular, they give a partial converse to Lagrange's Theorem, in that they show that for maximal prime power divisors of the order of a group G, there are subgroups of G of these orders.

Definition 11.1 *Let p be a prime number, let n be a positive integer and let k be an integer not divisible by p. Let G be a finite group with $p^n k$ elements. A* Sylow p-subgroup *is a subgroup of G with p^n elements.*

Note that for a general divisor d of $p^n k$, there is no reason why G should have a subgroup with d elements. The fact that G does have subgroups with p^n elements is established in Sylow's First Theorem, which we shall prove after the following preliminary lemma.

Lemma 11.2 *Let p be a prime integer and k be an integer not divisible by p. The number of ways of selecting a subset with p^n elements from a set with $p^n k$ elements is congruent to k modulo p.*

Proof It is well known that the number of ways of selecting a subset with p^n elements from a set with $p^n k$ elements is equal to the coefficient of x^{p^n} in the binomial expansion of

$$(1+x)^{p^n k} = ((1+x)^{p^n})^k.$$

However

$$(1+x)^{p^n} \equiv 1 + x^{p^n} \mod p,$$

since the coefficients of all other powers of x are divisible by p. Thus, the required coefficient of x^{p^n} is congruent modulo p to the coefficient of x^{p^n} in $(1+x^{p^n})^k$, and so is congruent to k modulo p. □

Example 11.3 The power of the Sylow theory will be illustrated as it is developed, by considering the problem of obtaining a complete list of the isomorphism types of groups with 15 elements. At the moment, we have very little useful information about such a group. For any integer n, there is always a cyclic group with n elements. So, we can think of an example of a group with 15 elements, namely the cyclic group C_{15}. How would we go about deciding whether or not this is the only group with 15 elements, up to isomorphism? We shall return to this question later.

Theorem 11.4 (The First Sylow Theorem) *Let p be a prime and G be a finite group of order kp^n, where p does not divide k. Then G has at least one Sylow p-subgroup.*

Proof The proof is an application of the Orbit–Stabiliser Theorem. Let $\mathcal{S}$ be the set of all subsets of G with p^n elements, so that, by Lemma 11.2, the number of elements in $\mathcal{S}$ is congruent to k modulo p. Make $\mathcal{S}$ into a G-set by defining, for any $S \in \mathcal{S}$, and any $g \in G$, $g \cdot S$ to be the set $gS = \{gs : s \in S\}$. Let $S_1, S_2, \ldots, S_r$ be representatives of the orbits of $\mathcal{S}$ under this action. Since each member of $\mathcal{S}$ is in the orbit of precisely one of the sets $S_1, \ldots, S_r$, we have a disjoint union

$$\mathcal{S} = \text{orb}(S_1) \cup \text{orb}(S_2) \cup \cdots \cup \text{orb}(S_r).$$

If each orbit had length divisible by p, then the total number of elements in $\mathcal{S}$ would be divisible by p. It follows by Lemma 11.2 that there must be at least one orbit, say that of S, such that the number of elements, m, in this orbit is not divisible by p. By the Orbit–Stabiliser Theorem (Theorem 10.16), the number of elements in the stabiliser G_S is of the form kp^n/m with m not divisible by p. Thus $|G_S|$ is of the form tp^n.

For any $g \in G_S, gS = S$, and so for any $s \in S, gs \in S$. It follows that the coset $G_S s$ is a subset of S. Thus

$$|G_S| = |G_S s| \leq |S| = p^n.$$

Since $|G_S|$ is of the form tp^n and is at most p^n, we deduce that G_S has precisely p^n elements. We have therefore established the existence of a Sylow p-subgroup of G, and we have also shown that for any element s of S, $G_S s = S$. □

Example 11.5 We now have some positive information about our group G of order 15. Since $15 = 3 \times 5$, this group will have at least one Sylow 3-subgroup (a group with 3 elements) and at least one Sylow 5-subgroup (a group with 5 elements). Furthermore any Sylow 3-subgroup is cyclic (being of prime order), as is any Sylow 5-subgroup. Note that it is only a coincidence that the Sylow p-subgroups have order p in this case. If the group G had 45 elements, the Sylow 3-subgroups would have order 9.

Theorem 11.6 (The Second Sylow Theorem) *The number of Sylow p-subgroups in a finite group G is congruent to 1 modulo p.*

Proof We consider the same G-set $\mathcal{S}$ as in the proof of Theorem 11.4. As we saw there, if the orbit in $\mathcal{S}$ of a set S has length not divisible by p, then for any $s \in S, G_S s = S$ so that $G_S = Ss^{-1}$ is a Sylow p-subgroup of G. It follows that $s^{-1}(G_S)s$ is a subgroup of G with the same number of elements as G_S, and so

$$s^{-1}(G_S)s = s^{-1}(Ss^{-1})s = s^{-1}S$$

is a Sylow p-subgroup in the orbit of S. Notice that the orbit of this subgroup (which is also the orbit of S) has k elements since $|G_S| = p^n$. Thus, if an orbit has length not divisible by p, then this orbit contains a Sylow p-subgroup and the length of the orbit is k.

Conversely, if an orbit contains a Sylow p-subgroup, P, say, then if g is in the stabiliser G_P, it follows that $gP = P$. Hence $g = g1 \in P$, so that $G_P \subseteq P$. Since P is contained in G_P, we deduce that $G_P = P$. The orbit of P therefore has length not divisible by p (in fact its length is k). Thus every orbit which contains a Sylow p-subgroup has length k, and every orbit of length not divisible by p contains a Sylow p-subgroup. Distinct Sylow p-subgroups are in distinct orbits, since if P_1 and P_2 are Sylow p-subgroups in the same orbit, P_1 would be equal to the left coset gP_2 for some $g \in G$. It would then follow that $1 \in P_2 \cap gP_2$, so that $P_2 = gP_2 = P_1$ by Corollary 5.5.

Thus, if n_p denotes the number of Sylow p-subgroups of G, we have shown that $|\mathcal{S}| \equiv kn_p$ modulo p. It now follows by Lemma 11.2 that $k \equiv kn_p$ modulo p, and so the number of Sylow p-subgroups is congruent to 1 modulo p, as required. □

Example 11.7 Theorem 11.6 shows that in a group of order 15, the number of Sylow 3-subgroups is one of $1, 4, 7, \ldots$ and the number of Sylow 5-subgroups is one of $1, 6, 11, \ldots$

Definition 11.8 *Let p be a prime integer. A p-group is a group in which every element has order a power of p. If G is a finite p-group, the First Sylow Theorem shows that the number of elements in G must be a power of p. Conversely, by Corollary 5.12, any finite group whose order is a power of p is a p-group.*

We need a preliminary before stating the next Sylow result.

Proposition 11.9 *Let P be a Sylow p-subgroup of a finite group G. Any p-subgroup of $N_G(P)$ is contained in P and in particular, P is the unique Sylow p-subgroup of $N_G(P)$.*

Proof Let Q be a p-subgroup of $N = N_G(P)$ and suppose that Q has order p^m, and that P has order p^n. Since P is a normal subgroup of N, by Proposition 7.10, $\langle P, Q\rangle = PQ$. Thus by Proposition 5.18, PQ is a subgroup of G of order p^{n+m-s}, where $|P \cap Q| = p^s$. Since p^n is the highest power of p dividing $|G|$, this is only possible if $m \leq s$. Since $P \cap Q$ is a subgroup of $Q, s \leq m$, and so we conclude that $s = m$ and $P \cap Q = Q$. It follows that Q is contained in P, as claimed. In particular, if Q is a Sylow p-subgroup of $N_G(P)$, then $Q = P$. □

Theorem 11.10 (The Third Sylow Theorem) *If P is a Sylow p-subgroup of the finite group G and Q is any p-subgroup of G, then Q is contained in a conjugate of P.*

Proof We consider the set $\mathcal{P}$ of distinct G-conjugates of P:

$$\mathcal{P} = \{gPg^{-1} : g \in G\}.$$

Consider the orbits of $\mathcal{P}$ under conjugation by P. The orbit of P is precisely $\{P\}$. We now show that P is the only element of $\mathcal{P}$ whose orbit is of length 1. If gPg^{-1} has one element in its orbit, then for all x in P, $x(gPg^{-1})x^{-1} = gPg^{-1}$. Thus for all x in P, $g^{-1}xg$ is an element in $N_G(P)$ whose order is the same as the order of x. It follows that $g^{-1}Pg$ is a p-subgroup of $N_G(P)$. In fact $P_1 = g^{-1}Pg$ is a Sylow p-subgroup of $N_G(P)$, since P and P_1 have the same number of elements. Hence $P_1 = P$ by Proposition 11.9, so that $gPg^{-1} = P$. This proves our claim that P is the only element of $\mathcal{P}$ whose orbit has length 1. We may therefore conclude that for any g not in P, the length of the orbit of gPg^{-1} under conjugation by elements of P has order greater than 1. By the Orbit–Stabiliser Theorem, these orbit lengths are all congruent to 0 modulo p. We deduce that $|\mathcal{P}| \equiv 1 \bmod p$.

Next consider the orbits of $\mathcal{P}$ under conjugation by elements of Q. Since every orbit has length a power of p, the above conclusion shows that there is at least one orbit of length 1, and so there is an element g such that for all x in Q, $x(gPg^{-1})x^{-1} = gPg^{-1}$. As in the previous paragraph, we deduce from this that $g^{-1}Qg$ is in $N_G(P)$, and so by Proposition 11.9, $g^{-1}Qg \subseteq P$, so that $Q \subseteq gPg^{-1}$, as required. □

Remark During the proof of Theorem 11.10, we showed that the number of elements in the orbit of any given Sylow p-subgroup is congruent to 1 modulo p. We know by the Second Sylow result, Theorem 11.6, that the number of Sylow p-subgroup is also congruent to 1 modulo p. The next result explains these facts.

Corollary 11.11 (The Fourth Sylow Theorem) *Any Sylow p-subgroups of a finite group G are conjugate, so the number of Sylow p-subgroups divides $|G|$.*

Proof Suppose that P and Q are Sylow p-subgroups of G. By Theorem 11.10, Q is contained in a conjugate of P. Since P and Q have the same number of elements, this is only possible if Q is equal to a conjugate of P. By the Orbit–Stabiliser Theorem, the number of conjugates of P is equal to the index of the normaliser $N_G(P)$ of P in G. Thus, by Lagrange's Theorem, the number of Sylow p-subgroups divides $|G|$. □

Example 11.12 We first return to our example of groups with 15 elements. The number of Sylow 3-subgroups is now known to divide 15 and also to be congruent to 1 modulo 3, so this number is 1. Let P be the unique Sylow 3-subgroup. Similarly, the number of Sylow 5-subgroups divides 15 and is congruent to 1 modulo 5, and so there is also a unique Sylow 5-subgroup, Q, say. Now let x be any element of G of order 3. Then by Proposition 4.11, $\langle x \rangle$ has 3 elements and so is a Sylow 3-subgroup of G. It follows that $\langle x \rangle = P$, and so x must equal one of the two non-identity elements of P. Similarly if y is an element of G of order 5, then $\langle y \rangle$ must be Q, and y is one of the four non-identity elements of Q. Thus G has: one element of order 1; two elements of order 3, and four elements of order 5. What orders could the remaining eight elements of G have? A corollary of Lagrange's Theorem, Corollary 5.12, shows that the order of any one of these eight elements divides 15 and so is $1, 3, 5$ or 15. Since the elements of orders 1, 3 and 5 have already been accounted for, each of these eight elements must have order 15. This means that the fifteen powers of such an element are all distinct, and so G is cyclic. We have therefore shown that any group of order 15 is cyclic.

Remark Recall that any conjugate of a subgroup H is also a subgroup. Furthermore, if H has finite order then $|H| = |gHg^{-1}|$. It follows that any conjugate of a Sylow p-subgroup is also a Sylow p-subgroup. Thus, if G has precisely one Sylow p-subgroup, then this subgroup is normal. Conversely, if a Sylow p-subgroup is normal, then using Corollary 11.11, we see that there can only be one Sylow p-subgroup.

Example 11.13 We now apply the Sylow results to a group G with 6 elements. We first consider the case $p = 3$. The number of Sylow 3-subgroups divides 6 (and so is 1, 2, 3 or 6) and is congruent to 1 modulo 3 (and so is 1, 4, 7, ...). Thus G has a unique Sylow 3-subgroup P, of order 3, which is therefore normal by the previous remark. Since a group with 3 elements is cyclic (Proposition 5.19), we see that $P = \langle x \rangle$ for some x with $x^3 = 1$.

Since G also has Sylow 2-subgroups of order 2, G has an element y, say, with $y^2 = 1$.

Since P is a normal subgroup, $y^{-1}xy \in P$, and so $y^{-1}xy$ is 1, x or x^2. However, $y^{-1}xy$ cannot equal 1 unless $x = 1$, contrary to definition. We deduce that there are two possibilities:

Case 1. If $y^{-1}xy = x$, then $xy = yx$ and the powers of xy are:

$$xy;$$

$$(xy)^2 = x(yx)y = x(xy)y = x^2y^2 = x^2;$$

$$(xy)^3 = (xy)^2(xy) = x^2xy = x^3y = y;$$

$$(xy)^4 = (xy)(xy)^3 = xyy = xy^2 = x;$$

$$(xy)^5 = (xy)^4(xy) = xxy = x^2y.$$

It follows that xy has six different powers and so the group is cyclic.

Case 2. If $y^{-1}xy = x^{-1}$, the six elements $\{1, x, x^2, y, xy, xy^2\}$ are distinct and form a group. Thus, in this case

$$G = \langle x, y : x^3 = 1 = y^2, xy = yx^{-1} \rangle,$$

so that G is isomorphic to the dihedral group $D(3)$.

We considered the possible groups with six elements in Chapter 5. The main part of the discussion there was taken up by proving that such a group has an element of order 3 and one of order 2. Now that we have the Sylow theory available, we can deduce these facts easily.

We conclude this chapter with some elementary properties of Sylow p-subgroups.

Proposition 11.14 *Let G be a finite group. Let P be a Sylow p-subgroup of G and let N be a normal subgroup of G. Then*

(i) *$P \cap N$ is a Sylow p-subgroup of N; and*
(ii) *PN/N is a Sylow p-subgroup of G/N.*

Proof First notice that one way to show that a subgroup H of a group G is a Sylow p-subgroup of G is to check that H is a p-subgroup and also that the index of H in G is not divisible by p.

(i) Since N is a normal subgroup of G, Proposition 7.10 may be used to show that $\langle P, N \rangle = PN$. Since $P \cap N$ is a subgroup of P, its order is a power of p. By the opening remark, it only remains to show that $|N : P \cap N|$ is not divisible by p. By Proposition 5.18, $|N : P \cap N| = |PN : N|$. However, by Corollary 5.13,

$$|G : P| = |G : PN||PN : P|,$$

so that $|PN : P|$ divides $|G : P|$. It follows that $|PN : P|$ is not divisible by p, so that $P \cap N$ is a subgroup of N of index not divisible by p. Hence, $P \cap N$ is a Sylow p-subgroup of N.

(ii) Since $PN/N \cong P/(P \cap N), PN/N$ is a p-subgroup of G/N. As in the proof of (i), $|G : PN|$ is not divisible by p, and so PN/N is a Sylow p-subgroup of G/N. □

Summary for Chapter 11

The basic results of this chapter are the Sylow Theorems, which may be collected together as follows:

Let p be a prime and G be a finite group with kp^n elements, where n is a non-negative integer and k is a positive integer not divisible by p. Then

(1) G has subgroups of order p^n (known as Sylow p-subgroups);

(2) the number, n_p, of these subgroups is of the form $1 + tp$, for some integer t;

(3) any subgroup Q of G whose order is a power of p is contained in some conjugate of a fixed Sylow p-subgroup P, so that $Q \subseteq gPg^{-1}$ for some $g \in G$;

(4) any two Sylow p-subgroups are conjugate, so that n_p divides $|G|$.

There is no universally accepted way to number these results, so that different authors may number the constituent parts of Sylow's theory in different ways.

These basic results on finite groups are, as we shall see, of key importance in understanding the structure of finite groups. They were published in 1872 by Sylow, a Norwegian mathematician who was only appointed to a university Professorship after he had retired from schoolteaching at the age of 65. The proof given using the Orbit–Stabiliser Theorem is based on a proof published by Wielandt in 1959.

Exercises 11

1. What (if anything) do the Sylow theorems enable one to deduce about the number of Sylow p-subgroups in the following cases:

(a) $p = 7; |G| = 28$;
(b) $p = 2; |G| = 48$;
(c) $p = 2; |G| = 32$;
(d) $p = 2; |G| = 12$;
(e) $p = 3; |G| = 12$.

2. Let H_i be a subgroup of G_i $(i = 1, 2)$. Show that $H_1 \times H_2$ is a subgroup of $G_1 \times G_2$, and that this subgroup is normal if H_i is a normal subgroup of G_i $(i = 1, 2)$. If H_i is a Sylow p-subgroup of G_i $(i = 1, 2)$, show that $H_1 \times H_2$ is a Sylow p-subgroup of $G_1 \times G_2$. Finally, if H_i is the unique Sylow p-subgroup of G_i $(i = 1, 2)$, show that $H_1 \times H_2$ is the unique Sylow p-subgroup of $G_1 \times G_2$.

3. Give an example of a group G with a Sylow p-subgroup P and a subgroup H such that $P \cap H$ is not a Sylow p-subgroup of H.

12

Applications of Sylow theory

In this chapter we apply the Sylow Theorems to various classification problems.

(1) Groups of order 2p

Let p be an odd prime number, and let G be a group with $2p$ elements. We may apply the Sylow theory to the primes 2 and p in turn. Thus the number n_p of Sylow p-subgroups divides $2p$ by Corollary 11.11, and is congruent to $1 \bmod p$ by Theorem 11.6. Hence n_p is one of $1, 2, p$ or $2p$. Since p and $2p$ are divisible by p these are both congruent to $0 \bmod p$. Since 2 is smaller than p, 2 is not congruent to $1 \bmod p$, so we conclude that n_p is 1. Thus the Sylow p-subgroup, P, is a normal subgroup of G. [Alternatively, once we know that P exists, the fact that the index of P is 2 means that P is a normal subgroup of G and hence is the unique Sylow p-subgroup of G.] Since P has p elements, P is cyclic, say $P = \langle x \rangle$. We also know that G has at least one Sylow 2-subgroup, so there is an element y of order 2. The elements of G are therefore

$$\{1,\ x,\ \ldots,\ x^{p-1},\ y,\ yx,\ \ldots,\ yx^{p-1}\}.$$

Since P is normal, yxy^{-1} is an element of P and so is of the form x^i for some i. Thus, since $y^2 = 1$,

$$(yx)^2 = yxy^{-1}x = x^{i+1}.$$

This means that the even powers of yx are powers of x whereas the odd powers of yx are of the form yx^j, for some j. By Corollary 5.12, the order of yx divides $2p$, and so is one of $1, 2, p$ or $2p$. If $i \neq -1$ in the above equation, we see that yx is not of order 1 (it is not the identity element), it is not of order 2 (since $(yx)^2$ is equal to x^{i+1}) and not of order p (since its pth power is of the form yx^j for some j). Thus yx must have order $2p$, so the powers of yx include all elements of G and G is cyclic. This implies that G is abelian and so, in fact, $yx = xy$. We have shown that when p is an odd prime, a group with $2p$ elements is either cyclic or is of the form

$$\langle x, y : x^p = 1 = y^2 \text{ and } yx = x^{-1}y \rangle,$$

so that G is isomorphic to the dihedral group $D(p)$.

Note that when $p = 2, G$ is a group of order 4, and these groups were classified in Chapter 3: a group with four elements is isomorphic either to C_4 or to $C_2 \times C_2$.

(2) Groups of order 21

There are two preliminary results required before the discussion of groups with 21 elements.

Proposition 12.1 *Let p and q be primes with $p > q$. A group of order pq has a normal Sylow p-subgroup.*

Proof The divisors of pq are $1, p, q$ and pq. Of these p and pq have remainder 0 when divided by p, and q has remainder q when divided by p, since q is less than p. There can therefore be only one Sylow p-subgroup, and so this subgroup is normal. □

Proposition 12.2 *Let x, y be elements of a group G such that $xy = yx$. Then, for all integers k, $(xy)^k = x^k y^k$.*

Proof This result was proved for disjoint permutations when k is a positive integer in Proposition 9.8. The same proof holds here. The case when $k = 0$ is trivial, and the case when $k < 0$ follows easily. □

Now let G be a group with 21 elements. Proposition 12.1 shows that G has a unique Sylow 7-subgroup $P = \langle x : x^7 = 1 \rangle$, say, and an element y of order 3. Since P is a normal subgroup of G, $yxy^{-1} = x^i$ for some i with $0 \le i \le 6$. Thus

$$\begin{aligned} x &= y^3xy^{-3} = y^2(yxy^{-1})y^{-2} \\ &= y^2x^iy^{-2} = (y^2xy^{-2})^i \\ &= (yx^iy^{-1})^i = (yxy^{-1})^{i^2} = x^{i^3}. \end{aligned}$$

Hence $i^3 \equiv 1 \bmod 7$, so 7 divides $i^3 - 1$. Considering the seven possible values of i in turn, we see that the only solutions for i are $i \equiv 1, 2$ or $4 \bmod 7$. In the first case, when $yxy^{-1} = x$, we see that $xy = yx$. Using Proposition 12.2, we see that $(xy)^3 = x^3$ and $(xy)^7 = y$, so the order of xy, being a divisor of 21, must equal 21, so that G is cyclic.

The cases when $i \equiv 2$ or $4 \bmod 7$ yield isomorphic groups since if y is an element of order 3 for which $yxy^{-1} = x^2$, then $z = y^2$ is an element of order 3 for which $zxz^{-1} = x^4$. Thus there are, up to isomorphism, two groups with 21 elements, the cyclic group and that with presentation

$$\langle x, y : x^7 = 1 = y^3,\ yxy^{-1} = x^2 \rangle.$$

In order to show that there is a group with 21 elements with this presentation, consider the matrices with entries from $\mathbf{Z}_7$:

$$X = \begin{pmatrix} 1 & 1 \\ 0 & 1 \end{pmatrix},\ Y = \begin{pmatrix} 4 & 0 \\ 0 & 2 \end{pmatrix}.$$

It is easily checked that these matrices satisfy the relations for the group of order 21.

(3) Groups of order 12

In many situations the Sylow results are used as the starting point of more detailed investigations. For example, we might be able to deduce that one or other Sylow subgroup is normal, as in the next few situations we consider. In general, these methods may not lead to a complete classification on their own. Note that we have now obtained some information on the structure of all groups with 11, or fewer, elements.

Proposition 12.3 *A group with 12 elements either has a normal Sylow 2-subgroup or a normal Sylow 3-subgroup.*

Proof Note that a Sylow 2-subgroup of the group G of order 12 has 4 elements. The number of Sylow 3-subgroups is either 1 or 4. We show that if this number is 4 then the number of Sylow 2-subgroups must be 1. If G has four distinct Sylow 3-subgroups P_1, P_2, P_3, P_4, each intersection $P_i \cap P_j$ $(i \neq j)$ is a proper subgroup of P_i, a group with three elements. It follows that $P_i \cap P_j = \{1\}$ for $i \neq j$. Thus G contains the identity element together with eight elements of order 3, two of these occurring in each of the four Sylow 3-subgroups. Only three elements remain, and so G has a unique Sylow 2-subgroup, this subgroup having three non-identity elements. □

Remark The detailed classification of groups with 12 elements will be given later in Chapter 22. The argument used to prove Proposition 12.3 is one layer more sophisticated than a simple counting argument, since it looked more closely at the possibility that the group had more than one Sylow 3-subgroup. Note that it is important to choose the primes in the correct order: if we had supposed that G had three Sylow 2-subgroups T_1, T_2 and T_3, we could not have concluded that $T_1 \cap T_2 = \{1\}$, since this intersection could contain two elements.

(4) Groups of order p^2q

The next case discussed is a general situation.

Proposition 12.4 *If p and q are distinct primes then a group of order p^2q has a normal Sylow subgroup.*

Proof The number of Sylow p-subgroups divides p^2q and is not a multiple of p, so is either 1 or q. If $p > q$, then q cannot be congruent to 1 mod p and so the number of Sylow p-subgroups is 1, as required. If, however, $q > p$, there could be q Sylow p-subgroups if $q \equiv 1 \bmod p$. In this case, the number of Sylow q-subgroups is not a multiple of q and divides p^2q, so it is $1, p$ or p^2. This number cannot be p (since p is not congruent to 1 mod q). If this number were p^2, we would have $p^2 \equiv 1 \bmod q$, so that q would divide $(p-1)(p+1)$. This can only occur if q divides $p-1$ or q divides $p+1$. However, $q > p$, so the only possibility is for q to equal $p+1$, which make p and q consecutive prime numbers and so p is 2 and q is 3. In this case, G has 12 elements, so the result then follows by Proposition 12.3. □

(5) Groups of order 30

We shall next consider groups of order 30. We first recall from Chapter 11 that any group with 15 elements is cyclic.

Proposition 12.5 *Let G be a group with 30 elements. Then G has a cyclic normal subgroup of order 15.*

Proof Applying Theorem 11.6 and Corollary 11.11, we see that G has one or ten Sylow 3-subgroups and also G has one or six Sylow 5-subgroups. The first step is to consider the possibility that there is a unique Sylow 3-subgroup P, say. Since P is normal, we may form the quotient group G/P, which has order 10. By the discussion in (1) above, G/P has a normal Sylow 5-subgroup N/P, say, in this case. By Theorem 7.14, N is a normal subgroup of G and N has 15 elements. Thus, by Example 11.12, N is a normal cyclic subgroup of G.

Now consider the possibility that G has ten distinct Sylow 3-subgroups $P_1, P_2, \ldots, P_{10}$, say. By Lagrange's Theorem, for $i, j \in \{1, 2, \ldots, 10\}$ and $i \neq j$, the number of elements in $P_i \cap P_j$ divides $|P_i| = 3$. Since P_i and P_j are distinct, we deduce that $P_i \cap P_j = \{1\}$. Thus each subgroup P_i contains the identity element together with two elements of order 3 which are not in any of the other subgroups P_j. This means that G has 20 distinct elements of order 3. There are therefore only $30 - (1 + 20) = 9$ elements of G of order different from 1 or 3. If G had six Sylow 5-subgroups, a similar argument to the above would show that there would then be 24 elements of order 5. This is impossible since only 9 elements of G were uncounted. This discussion shows that if G has 10 Sylow 3-subgroups, then it could only have one Sylow 5-subgroup, Q, say. Thus Q would be a normal subgroup of G and the quotient group G/Q would have order 6. By the discussion in (1), G/Q has a normal subgroup N/Q of order 3. Then Theorem 7.14 again shows that N is a normal subgroup of G of order 15. Thus if G has 10 Sylow 3-subgroups, then G also has a cyclic normal subgroup of order 15. □

Theorem 12.6 *A group of order 30 is either cyclic or dihedral or isomorphic to one of*

$$\langle x, y : x^{15} = 1 = y^2, yxy^{-1} = x^4 \rangle, \text{ or}$$
$$\langle x, y : x^{15} = 1 = y^2, yxy^{-1} = x^{11} \rangle.$$

Proof By Proposition 12.5, a group of order 30, G, has a normal cyclic subgroup N, say, of order 15. Let x be a generator for this subgroup, and y be a generator for any Sylow 2-subgroup of G. Since N is normal, $yxy^{-1} = x^i$ for some i. Then, since $y^2 = 1$,

$$x = y^2xy^{-2} = y(yxy^{-1})y^{-1} = yx^iy^{-1} = x^{i^2},$$

so that $i^2 - 1 \equiv 0 \bmod 15$. A case by case search shows that the only values of i which satisfy this congruence are $i \equiv 1, 4, 11$, or $14 \bmod 15$. The case $i \equiv 1 \bmod 15$ occurs precisely when G is cyclic and the case $i \equiv -1 \bmod 15$ precisely when G is dihedral. The other two cases yield the presentations

$$\langle x, y : x^{15} = 1 = y^2, yxy^{-1} = x^4 \rangle, \text{ and}$$
$$\langle x, y : x^{15} = 1 = y^2, yxy^{-1} = x^{11} \rangle,$$

as claimed. □

We shall show in the next chapter, in Theorem 13.8, that these two presentations do give groups of order 30, by showing that the groups are isomorphic to $C_3 \times D(5)$ and $C_5 \times D(3)$, respectively.

(6) Groups of order 24

As the final example of the use of Sylow's Theorems, we turn to a situation in which it is not even possible to ensure that a Sylow p-subgroup is normal.

Proposition 12.7 *Let G be a group with 24 elements. Then G has either a normal subgroup of order 8 or a normal subgroup of order 4.*

Proof The number of Sylow 2-subgroups is 1 or 3. If this number is 1, the Sylow 2-subgroup is a normal subgroup of order 8. We therefore suppose that G has three Sylow 2-subgroup S_1, S_2, S_3 each of which has order 8. By Proposition 5.18, the subset S_1S_2 has $2^3 2^3/2^r$ elements, where $|S_1 \cap S_2| = 2^r$. Since S_1S_2 is a subset of a group G with 24 elements, it follows that $2^3 2^3 \leq |S_1S_2| \times 2^r$ so that $64 = 2^6 \leq 24 \times 2^r$. Thus, $r \geq 2$. Since $S_1 \cap S_2$ is a proper subgroup of S_1, it has at most 2^2 elements, and so we deduce that if G has three Sylow 2-subgroups, then the intersection of any two of them has order 4.

Let $T = S_1 \cap S_2$, so that T has 4 elements. Since T is a subgroup of S_1 of index 2, T is a normal subgroup of S_1. Similarly, T is a normal subgroup of S_2. Thus S_1 and S_2 are both subgroups of $N_G(T)$, so $H = \langle S_1, S_2 \rangle$ is a subgroup of $N_T(G)$ and hence T is a normal subgroup of H. Since H is a subgroup, it contains S_1S_2. We have seen that S_1S_2 contains $2^6/2^2 = 16$ elements. Since the only subgroup of G containing at least 16 elements is G itself, we see that $H = G$ and so T is a normal subgroup of G of order 4. □

Summary for Chapter 12

Several applications of the Sylow results are given in this chapter. These lead to information about groups of orders $2p, pq, p^2q, 12, 24$ and 30. Among the complete classifications obtained are those for groups of order 15, $2p$ and 21.

Exercises 12

1. Prove that every group with 35 elements is cyclic.

2. Show that a group G with 105 elements either has a normal Sylow 5-subgroup or has a normal Sylow 7-subgroup. Deduce that G has a cyclic normal subgroup N with $|G : N| = 3$.

3. Show that a group with 56 elements either has a unique Sylow 2-subgroup or has a unique Sylow 7-subgroup.

13

Direct products

We have already met the direct product construction in Example 10 of Chapter 1. Recall that, given a pair of groups G and H, the direct product $G \times H$ is the set of ordered pairs (g, h) with $g \in G$ and $h \in H$ under the multiplication

$$(g_1, h_1)(g_2, h_2) = (g_1 g_2, h_1 h_2).$$

The following elementary properties of direct products are easily checked.

Proposition 13.1 *Let G and H be any groups. Then*

(1) *$G \times H$ is abelian if and only if both G and H are abelian;*
(2) *$G \times H$ is isomorphic to $H \times G$;*
(3) *if G and H are both cyclic finite groups and their orders have no common divisor greater than 1, then $G \times H$ is cyclic.*

Proof (1) This proof follows directly from the definitions; the details of the proof are given in the solutions to Question 6 of Exercises 1.

(2) The required map

$$\vartheta : G \times H \to H \times G$$

is defined by $\vartheta(g, h) = (h, g)$. It is a routine calculation to check that ϑ is a group homomorphism and also that ϑ is bijective.

(3) Suppose that G has order n and that x is a generator for G, and also that H has order m and that y is a generator for H. If the element (x, y) has order k, then

$$(x, y)^k = 1 = (x^k, y^k).$$

It follows by Proposition 3.10 that n divides k (since $x^k = 1$), and also that m divides k (since $y^k = 1$). Since n and m have no common divisor greater than 1, it follows that the sets of prime divisors of n and m are disjoint. It follows from this, after writing each of n and m as a product of prime powers, that nm divides k. However, $(x, y)^{nm} = (x^{nm}, y^{nm}) = 1$, so by Proposition 3.10 again, k divides nm. Thus the order of (x, y) is nm, and so (x, y) is a generator for $G \times H$. □

Corollary 13.2 *Let $n_1, n_2, \ldots, n_s$ be any sequence of integers each of which is greater than 1, such that the greatest common divisor of any distinct pair n_i, n_j is 1. Let G_i be a cyclic group of order n_i $(1 \leq i \leq s)$. Then the group $G_1 \times G_2 \times \cdots \times G_s$ is cyclic of order $n_1 n_2 \ldots n_s$.*

Proof The proof is by induction on s. The case when $s = 1$ is trivial. Assuming the result holds for $s-1$, we deduce that $H = G_1 \times G_2 \times \cdots \times G_{s-1}$ is cyclic of order $n_1 n_2 \ldots n_{s-1}$. Now apply Proposition 13.1 to the group $H \times G_s$ to obtain the result. □

Remark 1 Taking $G = H = C_2$, it may be checked that every element of the group $G \times H$ has order 2. We see that the group $G \times H$ cannnot therefore be cyclic, because it does not have an element of order 4. It follows that the claim of Proposition 13.1 (iii) requires the hypothesis that the orders of G and H have no common divisor greater than 1.

Remark 2 It is a common error to think that subgroups of $G \times H$ are of the form $G_1 \times H_1$ for G_1 a subgroup of G and H_1 a subgroup of H. However, not every subgroup of a direct product is of this type. For example, in the group $C_2 \times C_2$ mentioned in Remark 1, let x be a generator for G and y be a generator for H. The subgroups of G are therefore $\{1\}$ and G, and those of H are $\{1\}$ and H. However, the two elements $(1, 1), (x, y)$ form a subgroup of $G \times H$ which is not a direct product of a subgroup of G with a subgroup of H.

It is important to note that the groups G and H are not subsets of the Cartesian product and so are not subgroups of $G \times H$. For this reason, the construction we have just considered is sometimes known as the *external direct product* of G and H. We now introduce a related concept.

Definition 13.3 *Let $\{G_i : i = 1, \ldots, n\}$ be subgroups of G. Then G is the* internal direct product *of $\{G_i : i = 1, \ldots, n\}$ if*

(1) *each G_i is a normal subgroup of G; and*

(2) *every element of G can be written uniquely in the form*

$$g = g_1 g_2 \cdots g_n \text{ with } g_i \in G_i \ (1 \leq i \leq n).$$

Example 13.4 It is a consequence of Proposition 13.1(iii) that $C_2 \times C_3$ is isomorphic to C_6. In fact, we can easily see that C_6 is also an internal direct product. If x is a generator for C_6, let G_1 be the subgroup generated by x^2 and G_2 be the subgroup generated by x^3. Then since every subgroup of an abelian group is normal, G_1 and G_2 are both normal subgroups of C_6 isomorphic to C_3 and C_2, respectively. We can also list the unique 'factorisations' of the elements of C_6 as follows:

$$1 = 1 \cdot 1;\ x = x^3 \cdot (x^2)^2;\ x^2 = 1 \cdot x^2;$$

$$x^3 = x^3 \cdot 1;\ x^4 = (x^2)^2;\ x^5 = x^3 \cdot x^2.$$

Proposition 13.5 *Let $G_1, G_2, \ldots, G_n$ be normal subgroups of G such that*

(a) $(G_1 \ldots G_{i-1}) \cap G_i = \{1\} \qquad (i = 2, \ldots, n)$; and
(b) $G = G_1 G_2 \ldots G_n$.

Then G is the internal direct product of the groups $G_1, G_2, \ldots, G_n$.

Proof Condition (b) ensures that each element g of G can be written in the form $g = g_1 g_2 \ldots g_n$ with $g_i \in G_i$. We must show that this expression is unique. Suppose therefore that

$$g = g_1 g_2 \ldots g_n = h_1 h_2 \ldots h_n,$$

where $g_i, h_i \in G_i$ for $1 \leq i \leq n$ and $g_j \neq h_j$ for some j. Let j be the largest integer such that $g_j \neq h_j$, so that $g_i = h_i$ for $i > j$. Cancelling g_i for $i > j$, we see that $g_1 \ldots g_j = h_1 \ldots h_j$ and so

$$g_j h_j^{-1} = (g_1 \ldots g_{j-1})^{-1}(h_1 \ldots h_{j-1}) \in (G_1 \ldots G_{j-1}) \cap G_j.$$

However,

$$(G_1 \ldots G_{j-1}) \cap G_j = \{1\},$$

by (b). Thus $h_j = g_j$ contrary to the definition of j. This shows that the decomposition is unique. □

Remark In the case when $n = 2$, it follows from Proposition 13.5 that the condition for G to be the internal direct product of G_1 and G_2 is that G_1 and G_2 are normal subgroups of G such that $G = G_1 G_2$ and $G_1 \cap G_2 = \{1\}$. It also follows that whenever G is the internal direct product of $G_1, \ldots, G_n$, then $G_i \cap G_j = \{1\}$ for distinct i and j.

Corollary 13.6 *Let G be an internal direct product of $G_1, G_2, \ldots, G_n$. If x and y are elements of G_i and G_j, respectively, with $i \neq j$, then $xy = yx$.*

Proof Consider the element $g = xyx^{-1}y^{-1}$. Since y is in the normal subgroup G_j, xyx^{-1} is also in G_j and so $g \in G_j$. Similarly, $x \in G_i$, and so since G_i is a normal subgroup of G, $yx^{-1}y^{-1} \in G_i$ and hence $g \in G_i$. We have shown that $g \in G_i \cap G_j$. By the above remark, $G_i \cap G_j = \{1\}$, so that $g = 1$ and $xy = yx$, as required. □

We now investigate the relationship between the two forms of direct product.

Proposition 13.7 *Suppose that G is the external direct product of groups $G_1, G_2, \ldots, G_n$. Then there are normal subgroups $N_1, N_2, \ldots, N_n$ of G with N_i isomorphic to G_i (for $1 \leq i \leq n$) such that G is the internal direct product of $N_1, N_2, \ldots, N_n$. Conversely, if G is the internal direct product of $N_1, N_2, \ldots, N_n$, then G is isomorphic to the external direct product of groups isomorphic to $N_1, N_2, \ldots, N_n$.*

Proof Suppose that G is an external direct product of $G_1, G_2, \ldots, G_n$. For $1 \leq i \leq n$, define N_i to be the set

$$\{1\} \times \cdots \times \{1\} \times G_i \times \{1\} \times \cdots \times \{1\}$$

of elements which have entry 1 everywhere except possibly in the ith coordinate. It is then easily checked that

(i) N_i is isomorphic to G_i;
(ii) N_i is a normal subgroup of G; and
(iii) every element of G has a unique expression

$$(g_1, \ldots, g_n) = (g_1, 1, \ldots, 1)(1, g_2, \ldots, 1) \ldots (1, \ldots, g_n)$$

as a product of elements of $N_1, \ldots, N_n$.

Conversely, if G is the internal direct product of $N_1, \ldots, N_n$, define a map $\vartheta : G \to N_1 \times \cdots \times N_n$ by

$$\vartheta(g_1 g_2 \cdots g_n) = (g_1, g_2, \ldots, g_n).$$

A straightforward inductive proof using Corollary 13.6 shows that

$$\begin{aligned} \vartheta(g_1 g_2 \ldots g_n)\vartheta(h_1 h_2 \ldots h_n) &= \vartheta(g_1 h_1 g_2 h_2 \ldots g_n h_n) \\ &= \vartheta(g_1 \ldots g_n h_1 \ldots h_n), \end{aligned}$$

so that ϑ is a homomorphism. The fact that ϑ is bijective follows easily from the definitions. □

We now return to the classification of groups of order 30 which was started in Chapter 12.

Theorem 13.8 *A group of order 30 is isomorphic to one of*

$$C_{30},\ C_5 \times D(3),\ C_3 \times D(5),\ \text{or } D(15).$$

Proof By Theorem 12.6, we only need to show why the presentations

$$\langle x, y : x^{15} = 1 = y^2, yxy^{-1} = x^{11} \rangle \text{ and}$$

$$\langle x, y : x^{15} = 1 = y^2, yxy^{-1} = x^4 \rangle$$

give the groups $C_5 \times D(3)$ and $C_3 \times D(5)$, respectively. In the first case, let z denote x^3. Then

$$yzy^{-1} = yx^3y^{-1} = (yxy^{-1})^3 = (x^{11})^3 = x^{33} = x^3 = z$$

so that z commutes with y. Since z is a power of x, the element z also commutes with x and so z is in the centre of G. It follows that $\langle z \rangle$ is a normal subgroup of G of order 5.

Now let K be the subgroup of G generated by x^5 and y. Note that since

$$yx^5y^{-1} = (yxy^{-1})^5 = (x^{11})^5 = x^{55} = x^{10} = x^{-5},$$

the generators of K satisfy the relations

$$(x^5)^3 = 1 = y^2, \qquad yx^5y = x^{-5}.$$

Writing w for x^5 we see that K is generated by w and y, where

$$w^3 = 1 = y^2 \text{ and } wy = yw^2,$$

so that K is isomorphic to the dihedral group $D(3)$. It now follows by Proposition 13.7 that we only need to show that G is an internal direct product of K and $\langle z \rangle$. To do this, we use Proposition 13.5, and since $K \cap \langle z \rangle = \{1\}$, we only need to prove that K is a normal subgroup of G. Since K has 6 elements, and K is a subgroup of its normaliser $N_G(K)$, the order of $N_G(K)$ is divisible by 6 . By Lagrange's Theorem, $|N_G(K)|$ divides 30. If we show that x normalises K, it will then follow that the normaliser has order greater than 6, so that the normaliser would be the whole of G. It would then follow that K is a normal subgroup of G. To show that x normalises K, note that

$$xx^5x^{-1} = x, \text{ and } xyx^{-1} = xy^{-1}x^{-1} = yx^{11}x^{-1} = yx^{10} \in N.$$

We have therefore shown that K and $\langle z \rangle$ are normal subgroups of G with $K \cap \langle z \rangle = \{1\}$. It easily can be seen that $G = K\langle z \rangle$ so that, by Proposition 13.5, G is an internal direct product of K and $\langle z \rangle$.

Note that the question as to whether a group with this presentation really has 30 elements is answered by considering the group $C_5 \times D(3)$.

The argument in the other case is similar and is left to the reader as an exercise. □

To conclude this section, we discuss two useful constructions associated with direct products.

Definition 13.9 *Let G and H be groups with Z a subgroup of the centre of G and let W be a subgroup of the centre of H. Suppose that there is an isomorphism $\vartheta : Z \to W$. Using elementary properties of homomorphisms, it may be seen that the set $X = \{(x, \vartheta(x)^{-1}) : x \in Z\}$ is a subgroup of the direct product $G \times H$. In fact, since Z and W are central, it is easily seen that X is a central subgroup of the direct product $G \times H$. The quotient group $(G \times H)/X$, denoted $G \times_\vartheta H$, is the* central product *of G and H via ϑ.*

Example 13.10 Let G be the dihedral group of order 8 generated by a of order 4 and b of order 2, so that the centre of G is $\{1, a^2\}$. Let H be the quaternion group of order 8 generated by x and y so that the centre of H is $\{1, x^2\}$. Every element of H not in its centre has order 4 and square equal to a^2. Let $Z = \{1, a^2\}$ and $W = \{1, z^2\}$. Define a map $\vartheta : Z \to W$ by

$$\vartheta(1) = 1 \text{ and } \vartheta(a^2) = x^2.$$

Thus, the set X, as defined in 13.9, consists of $\{(1,1), (a^2, x^2)\}$. Since the group $G \times H$ has 64 elements, the central product has 32 elements. In the direct product:

the element $(1,1)$ has order 1;
the squares of the elements $(a^2, 1)$ and $(1, x^2)$ are in X;
the squares of the elements (a, h) (for h of order 4) are in X;
the squares of the elements (a^3, h) (for h of order 4) are in X;
the squares of the elements $(g, 1)$ (for g of order 2) are in X;
the squares of the elements (g, x^2) (for g of order 2) are in X;
all remaining elements of $G \times H$ are of order 4 and none of their squares is in X.

Thus there are 24 elements of the direct product, each of which has an image of order dividing 2 in the central product. It follows that these 24 elements are 12 cosets of X (including the subgroup X itself). Thus the central product has an identity element and 11 elements of order 2. The remaining 20 elements of the central product all have order 4.

In a similar way, it is possible to form a central product with G and H both equal to $D(4)$ by taking $Z = W = \{1, a^2\}$. This central product has 19 elements of order 2. It follows that the central product of $D(4)$ with Q_8 is not isomorphic to the central product of $D(4)$ with itself.

We now come to the other general construction.

Definition 13.11 *Let G and H be groups which have isomorphic quotient groups, so that there are normal subgroups N of G and K of H such that there is an isomorphism $\vartheta : G/N \to H/K$.* The pullback $G \times^{\vartheta} H$ of G and H via ϑ *is the subset of $G \times H$ of elements of the form (g, h), where $\vartheta(gN) = hK$. It is easily checked that the pullback is a subgroup of $G \times H$.*

Example 13.12 Let G and H be groups with normal subgroups N and K, respectively, each of index 2. The pullback in this case consists of those pairs (g, h) for which *either* $g \in N$ and $h \in K$ *or* $g \notin H$ and $h \notin K$. In particular, if G and H are symmetric groups, we can take the subgroups as the alternating subgroups in G and H. The pullback is then the permutations of the form (π, ρ) where π and ρ are either both even or both odd.

Summary for Chapter 13

In this chapter, we return to the idea of a direct product of two groups. It is now convenient to refer to the construction $G \times H$ given in Chapter 1 as the *external direct product* of G and H, since the groups G and H are not subsets of $G \times H$. By contrast, the *internal direct product* is constructed from subgroups of a group. We say that G is an internal direct product of H and N if H and N are normal subgroups of G with each element of G uniquely of the form hn, where $h \in H$ and $n \in N$. It is shown that the two forms of direct product are closely related, so that a group which is an internal direct product of subgroups $H_1, H_2, \ldots, H_r$ is isomorphic to an external direct product of groups which are isomorphic to $H_1, H_2, \ldots, H_r$, and conversely.

The internal direct product construction is used to complete the description of groups with 30 elements. The chapter ends with two constructions which arise from direct products: the central product of G and H and the pullback of G and H. These constructions provide a useful way to extend our catalogue of groups.

Exercises 13

1. Show that the dihedral group $D(4)$ is not an internal direct product of any two of its proper subgroups.

2. Prove that the group with presentation

$$\langle x, y : x^{15} = 1 = y^2,\ yxy^{-1} = x^4 \rangle$$

is isomorphic to the direct product $C_3 \times D(5)$.

3. Show that the dihedral group $D(6)$ is an internal direct product of a subgroup of order 2 with a subgroup of order 6 isomorphic to the dihedral group $D(3)$.

4. Complete the calculation of the number of elements of order 2 in the central product of two copies of the group $D(4)$. Find how many elements of order 2 there are in the central product of two copies of the quaternion group of order eight.

5. Let G be a cyclic group of order 6 generated by x and H be the alternating group $A(4)$, with N equal to $\langle x^3 \rangle$ and K equal to the subgroup $\{1, (1\ 2)(3\ 4), (1\ 3)(2\ 4), (1\ 4)(2\ 3)\}$. Define a map $\vartheta : G/N \to H/K$ by

$$\vartheta(N) = K, \quad \vartheta(xN) = (1\ 2\ 3)K \quad \text{and } \vartheta(x^2N) = (1\ 3\ 2)K.$$

List the elements in the pullback of G and H via ϑ.

14

The classification of finite abelian groups

In this chapter, we shall solve the classification problem for finite abelian groups. This is done by producing a 'standard' list of finite abelian groups with the property that each finite abelian group is isomorphic to precisely one standard group. The list of standard finite abelian groups with n elements is easy to describe. There is precisely one group with one element. For each integer n greater than 1, and each sequence of positive integers $n_1, n_2, \ldots, n_r$ such that n_1 is divisible by n_2, n_2 is divisible by $n_3, \ldots, n_{r-1}$ is divisible by n_r, and $n = n_1 n_2 \ldots n_r$ with $n_r \geq 2$, the standard group corresponding to this sequence is the direct product

$$C_{n_1} \times C_{n_2} \times \cdots \times C_{n_r}.$$

It will be proved in Corollary 14.11 below that every finite abelian group is isomorphic to precisely one of these standard groups.

Example 14.1 Consider the case of groups of order 100. The possible sequences of positive integers are: $\{100\}$ (when $r = 1$); $\{50, 2\}$; $\{20, 5\}$ and $\{10, 10\}$ (when $r = 2$). The divisor restriction means that there is no sequence with $r = 3$. This gives four standard abelian groups of order 100:

$$C_{100};\ C_{50} \times C_2;\ C_{20} \times C_5;\ C_{10} \times C_{10}.$$

Of course, we have yet to show that this is the complete list of abelian groups with 100 elements.

The study of abelian groups is much easier than that of non-abelian groups. There are several reasons for this. For example, every subgroup of an abelian group is normal. Perhaps one of the most significant properties of an abelian group is that which enables powers of elements to be computed.

Proposition 14.2 *Let A be an abelian group, and let x and y be elements of A. Then for any positive integer k, $(xy)^k = x^k y^k$.*

Proof The proof follows directly from Proposition 12.2, since G is abelian. □

The first step in the classification will be to consider the consequences of the Sylow theory for finite abelian groups. This will appear as a special case of a general result which we shall need later.

Proposition 14.3 *Let G be a finite group such that, for every prime p dividing $|G|$, the Sylow p-subgroup of G is normal. Let $p_1, p_2, \ldots, p_r$ be the distinct primes dividing $|G|$, and let P_i be the Sylow p_i-subgroup of G (for $1 \leq i \leq r$). Then G is the internal direct product*

$$P_1 \times P_2 \times \cdots \times P_r.$$

In particular, every finite abelian group is the direct product of its Sylow p-subgroups.

Proof Since each P_i is a normal subgroup of G, it follows by Proposition 7.10 that for $s \leq r$,

$$\langle P_1, P_2, \ldots, P_s \rangle = P_1 P_2 \ldots P_s,$$

so that $P_1 P_2 \ldots P_s$ is a subgroup of G. We now show by induction on s that for $s \leq r, P_1 P_2 \ldots P_s$ is of order $|P_1||P_2|\ldots|P_s|$. The result is clear when $s = 1$. By Proposition 5.18,

$$|(P_1 P_2 \ldots P_{s-1}) P_s| = |P_1 P_2 \ldots P_{s-1}||P_s| / |(P_1 P_2 \ldots P_{s-1}) \cap P_s|.$$

By induction

$$|P_1 P_2 \ldots P_{s-1}| = |P_1||P_2|\ldots|P_{s-1}|$$

and so, since p_s does not divide $|P_1 P_2 \ldots P_{s-1}|$, we deduce that

$$|(P_1 P_2 \ldots P_{s-1}) \cap P_s| = 1$$

It follows that

$$|P_1 P_2 \ldots P_{s-1} P_s| = |P_1||P_2|\ldots|P_s|,$$

as claimed.

Applying this with $s = r$ shows that $P_1 P_2 \ldots P_r$ is a subgroup of G with $|G|$ elements, so that this subgroup is equal to G. We have also seen that

$$(P_1 P_2 \ldots P_{s-1}) \cap P_s = \{1\}, \quad (2 \leq s \leq r).$$

It therefore follows by Proposition 13.5 that G is the internal direct product of $P_1, P_2, \ldots, P_r$.

When G is abelian, the result follows because every subgroup of an abelian group is normal. □

This result reduces the study of finite abelian groups to finite abelian p-groups.

Example 14.4 We have already seen that for any prime p, a group of order p^2 is abelian and is either cyclic of order p^2 or $C_p \times C_p$. Thus, if G is a finite abelian group of order $100 = 2^2 5^2$, we see that $G = P \times Q$, where P has order 4 and Q has order 25. This gives us four possible types of abelian groups with 100 elements:

$$C_4 \times C_{25};\ C_2 \times C_2 \times C_{25};\ C_4 \times C_5 \times C_5;\ \text{and } C_2 \times C_2 \times C_5 \times C_5.$$

Using Proposition 13.1, we see that $C_4 \times C_{25}$ is isomorphic to C_{100}; that $C_2 \times C_2 \times C_{25}$ is isomorphic to $C_{50} \times C_2$; that $C_4 \times C_5 \times C_5$ is isomorphic to $C_{20} \times C_5$; and that $C_2 \times C_2 \times C_5 \times C_5$ is isomorphic to $C_{10} \times C_{10}$. We have therefore established that the list of standard abelian groups is the complete list of groups with 100 elements. This depended on the fact that we already know the classification of (abelian) p-groups of order p^2.

We now give the existence part of the classification of finite abelian groups.

Proposition 14.5 *Let G be a finite abelian p-group of order p^n. Then G is an internal direct product of cyclic subgroups of orders $p^{e_1}, p^{e_2}, \ldots, p^{e_r}$, where $e_1 \geq e_2 \geq \cdots \geq e_r \geq 1$ and $e_1 + e_2 + \cdots + e_r = n$.*

Proof The proof is by induction on n, the result being clear when $n = 1$. Choose an element x of G of maximal possible order among all the elements of G, and let A be the subgroup of G generated by x. The objective of the proof is to show that there is a subgroup B of G such that G is the internal direct product of A and B. Once this is done, we may apply the inductive hypothesis to B, to see that G is an internal direct product of cyclic subgroups. Since the order in which these subgroups occur can be altered using Proposition 13.1(ii), the result will then follow.

There is a preliminary needed before we can define B. By induction, G/A is an internal direct product of cyclic groups, generated by the cosets $\langle y_1 A\rangle, \langle y_2 A\rangle, \ldots, \langle y_s A\rangle$ of orders $p^{m_1}, p^{m_2}, \ldots, p^{m_s}$, say. For $1 \leq i \leq s$, we therefore have that $(y_i)^{p^{m_i}} = x^{t_i}$ for some t_i. Suppose that for some value of i, p^{m_i} does not divide t_i, so that $t_i = k_i p^{r_i}$, where p does not divide k_i and $r_i < m_i$. Proposition 4.15 then shows that x^{k_i} generates $\langle x\rangle$, and so we see that x^{t_i} has order p^{m-r_i}, where p^m is the order of x. It then follows that the order of y_i is $p^{m-r_i+m_i}$ so that y_i has order greater than the order of x, contrary to the definition of x. We may therefore suppose that p^{m_i} divides t_i, and so $(y_i)^{p^{m_i}} = (x_i)^{p^{m_i}}$ for some $x_i = x^{t_i/p^{m_i}} \in A$ for $1 \leq i \leq s$.

Now write $z_i = y_i x_i^{-1} (1 \leq i \leq s)$ and let B be the group generated by $z_1, z_2, \ldots, z_s$. Note that, since $(z_i)^{p^{m_i}} = 1$, the order of z_i divides p^{m_i}.

However, $z_iA = y_iA$, and y_iA has order p^{m_i} in G/A, so the order of z_i cannot be less than p^{m_i} and so is precisely p^{m_i}. We shall show that G is the internal direct product of A and B by showing that

(i) every element of G can be written in the form ab, with $a \in A$ and $b \in B$, and

(ii) $A \cap B = \{1\}$.

To establish the first of these claims, take any element g of G. Since the cosets $z_1A, z_2A, \ldots, z_sA$ are equal to $y_1A, y_2A, \ldots, y_sA$, they generate the group G/A. The coset gA can therefore be written as a product of powers of the cosets $z_1A, z_2A, \ldots, z_sA$. It follows that g is a product of an element of A with an element of B. To establish the second claim, suppose that $a \in A \cap B$, so that

$$a = z_1^{k_1} z_2^{k_2} \ldots z_s^{k_s}$$

and so

$$aA = z_1^{k_1} A \, z_2^{k_2} A \ldots z_s^{k_s} A = y_1^{k_1} A \, y_2^{k_2} A \ldots y_s^{k_s} A.$$

Since a is in A, this coset is A. Since $y_1A, y_2A, \ldots, y_sA$ generate G/A, and G/A is isomorphic to the direct product

$$C_{p^{m_1}} \times \cdots \times C_{p^{m_s}},$$

this means that p^{m_i} divides k_i $(1 \leq i \leq s)$, and so $a = 1$, as required. This completes the proof. □

This shows that every finite abelian p-group is isomorphic to one of the standard list of p-groups discussed at the beginning of this section.

Example 14.6 When $p = 2$, we see that the list of abelian 2-groups of order 8 is $C_8, C_4 \times C_2$ and $C_2 \times C_2 \times C_2$. The list of abelian groups of order 16 is

$$C_{16},\ C_8 \times C_2,\ C_4 \times C_4,\ C_4 \times C_2 \times C_2 \text{ and } C_2 \times C_2 \times C_2 \times C_2.$$

The process illustrated in Example 14.4 can be generalised for any integer n to give the following result.

Corollary 14.7 *A finite abelian group of order n can be written as a direct product*

$$C_{n_1} \times C_{n_2} \times \cdots \times C_{n_r},$$

where n_i is divisible by n_j for $j > i, n_r \geq 2$, and $n_1 n_2 \ldots n_r = n$.

Proof First use Proposition 14.3 to write G as an internal direct product of its Sylow subgroups, then use Proposition 14.5 to write each Sylow subgroup as a direct product of cyclic groups. Now use Corollary 13.2 repeatedly on the first factor of each Sylow subgroup to obtain a cyclic factor of G. Then use Corollary 13.2 again on the second factors of those Sylow subgroups which are not cyclic (if there are any) to obtain a second cyclic factor of G. Continue in this way to complete the proof. □

We are now left with the question of uniqueness: is it possible that two 'standard' finite abelian groups could be isomorphic? We first consider the situation for abelian p-groups.

Definition 14.8 *In order to avoid excessive use of subscripts and superscripts, we say that a finite abelian p-group G is of* type $(e_1, e_2, \ldots, e_r)$ *if*

$$G \cong C_{p^{e_1}} \times C_{p^{e_2}} \times \cdots \times C_{p^{e_r}} \text{ and } e_1 \geq e_2 \geq \cdots \geq e_r \geq 1.$$

The next result introduces two standard subgroups of a finite abelian group.

Proposition 14.9 *Let p be a prime. For any finite abelian group G, the subset G_p of G consisting of elements of order 1 or p is a subgroup of G. Also the subset G^p of p-th powers of elements of G is a subgroup of G. Let G be an abelian p-group of type $(e_1, e_2, \ldots, e_r)$, with t the largest integer such that $e_t > 1$. Then G_p has order p^r, and G^p has type $(e_1-1, e_2-1, \ldots, e_t-1)$.*

Proof If x and y are in G_p, then $(xy)^p = x^p y^p$ by Proposition 14.2, so xy has order 1 or p. Since the inverse of an element of order p also has order p, we see that G_p is a subgroup of G. Similarly, the pth powers of elements of G form a subgroup because $x^p y^p = (xy)^p$, and the inverse of a pth power is also a pth power. The claims about the order of G_p and the type of G^p are easily checked since G^p is generated by $x_1^p, x_2^p, \ldots, x_r^p$, where $x_1, x_2, \ldots, x_r$ are generators for G. □

This result is now used to show that the type of G is unique.

Proposition 14.10 *Suppose that G is an abelian p-group. Then the type of G is uniquely determined. Thus, if G is of type $(e_1, e_2, \ldots, e_r)$ and also of type $(f_1, f_2, \ldots, f_s)$ then $r = s$, and $e_i = f_i$ for $1 \leq i \leq r$.*

Proof The proof is by induction on $|G|$. If G has type $(e_1, e_2, \ldots, e_r)$, then it follows by Proposition 14.9 that G_p has order p^r. Thus $r = s$, and the result is proved if $G_p = G$. Now consider the subgroup G^p. If G is of type $(e_1, e_2, \ldots, e_r)$, applying Proposition 14.9 shows that G^p is of type $(e_1 - 1, e_2 - 1, \ldots, e_t - 1)$, where t is the largest integer such that $e_t > 1$. Applying induction to G^p gives the result. □

We now come to the classification theorem for finite abelian groups.

Corollary 14.11 *Every finite abelian group G has a unique decomposition in the form*

$$C_{n_1} \times C_{n_2} \times \cdots \times C_{n_r},$$

where n_i is divisible by n_j for $j > i, n_r \geq 2$, and

$$n_1 n_2 \ldots n_r = |G|.$$

Proof Corollary 14.7 gives the existence of such a decomposition for G. To prove uniqueness, note that for each prime p dividing $|G|$, the Sylow p-subgroup of G is

$$S_p(C_{n_1}) \times S_p(C_{n_2}) \times \cdots \times S_p(C_{n_r}),$$

where $S_p(K)$ denotes the Sylow p-subgroup of the abelian group K. Now apply Proposition 14.10 to this decomposition. □

As an application of the classification theorem, we consider the structure of finite fields.

Definition 14.12 *A* field *is a set F which is an abelian group under addition. There is also a multiplication on F which is closed, associative and commutative, and there is an identity element 1_F which is not equal to the zero element. The other axioms are the distributive laws:*

$$x(y+z) = xy + xz \text{ and } (x+y)z = xz + yz;$$

and the requirement that every non-identity element has a multiplicative inverse:

$$\text{if } x \neq 0 \text{ then there exists } y \in F \text{ such that } xy = 1_F.$$

Notice that the non-zero elements of a field F form a group under multiplication. It is an easy consequence of these axioms that $0x = 0$ for all elements x in a field F.

We have made almost no use of the additive notation for abelian groups, so it may be helpful to look at the interpretation of some of the basic ideas before we continue. The zero element 0 is the additive identity, and for any x in F, and positive integer n, we denote by nx the result of adding x together n times. Then the additive order of x is the smallest positive integer such that $nx = 0$. It follows by the Remark before Proposition 3.10

that if F is a field with a finite number of elements, then every element of F has finite additive order. Define the *characteristic of* F to be the additive order of 1_F. Let k be the characteristic of F, and suppose that d_1 and d_2 are proper divisors of k with $k = d_1 d_2$. Then by the additive version of the index laws,

$$0 = k1 = d_1 d_2 1 = (d_1 1)(d_2 1).$$

Since F has no zero divisors, this means that one of $d_1 1$ or $d_2 1$ must be zero. This contradicts the definition of k, so k has no proper divisors. We have therefore shown that the characteristic of F is a prime integer.

Definition 14.13 *A finite abelian p-group G is* elementary abelian *if every element of G has order dividing p. It follows from the classification theorem that an elementary abelian p-group is of the form* $C_p \times C_p \times \cdots \times C_p$, *so that G has type* $(1, 1, \ldots, 1)$.

Proposition 14.14 *Let F be a finite field. Then there is a prime p and a positive integer n such that F has* p^n *elements. The additive group of F is elementary abelian.*

Proof Let p be the characteristic of F. Now notice that for any x in F,

$$px = p(1x) = (p1)x = 0,$$

so every element of F has order p. It then follows by definition that F is elementary abelian, and so $|F| = p^n$ for some n. □

Proposition 14.15 *The multiplicative group of a finite field is cyclic.*

Proof Apply the classification theorem for finite abelian groups to the multiplicative group $F^* = F - \{0\}$. Thus this group is a direct product of cyclic groups

$$C_{n_1} \times C_{n_2} \times \cdots \times C_{n_r},$$

where n_1 divides n_2, n_2 divides $n_3, \ldots$, and n_{r-1} divides n_r. It follows that the order of each element of the group divides n_1, so that each element of F^* satisfies the equation $x^{n_1} = 1$. However, the polynomial $x^{n_1} - 1$ has at most n_1 roots in F since, if x_0 is a root of a polynomial $f(x)$, then $x - x_0$ divides $f(x)$, so that the number of roots of a polynomial is at most the degree of the polynomial. It follows that F^* has n_1 elements so that $n_1 = |F^*|$ and F^* is cyclic. □

Summary for Chapter 14

In this chapter, the classification of finite abelian groups is given. It is shown that every finite abelian group G has a unique decomposition of the form

$$C_{n_1} \times C_{n_2} \times \cdots \times C_{n_r},$$

where n_i is divisible by n_j for $j > i, n_r \geq 2$, and

$$n_1 n_2 \ldots n_r = |G|.$$

This means that the complete list of finite abelian groups of any given order can be written down. For example, the abelian groups of order 36 are C_{36}, $C_{18} \times C_2$, $C_{12} \times C_3$ and $C_6 \times C_6$.

As an illustration of the use of this result, it is shown that a finite field F has order p^n for some prime p and some positive integer n. Furthermore, the additive group of F is elementary abelian whereas the multiplicative group of non-zero elements of F is cyclic.

The classification of finite abelian groups was essentially obtained by Kronecker in 1870, although he did not prove uniqueness, and was concerned with a situation which arose in algebraic number theory rather than group theory.

Exercises 14

1. Write down the complete list of abelian groups with 360 elements.

2. Prove that in a finite abelian p-group the elements of order dividing p^r form a subgroup. Give an example of a finite p-group in which the elements of order dividing p do not form a subgroup.

3. Prove that if d is any divisor of the order of a finite abelian group G, then G has a subgroup of order d.

4. Determine how many subgroups of order p^2 there are in $C_p \times C_{p^2}$.

5. Determine the number of subgroups with p elements in the elementary abelian group with p^n elements. Find a non-elementary finite abelian p-group with $p^2 + p + 1$ subgroups of order p.

15

The Jordan–Hölder Theorem

In this chapter, we shall develop some of the basic results required to investigate chains of normal subgroups of a group G. The objective is to prove the Jordan–Hölder Theorem. This will be accomplished in three steps. We first introduce some basic definitions.

Definition 15.1 *Given a group G, a* subnormal series *for G is a chain*

$$G = G_0 \geq G_1 \geq G_2 \geq \cdots \geq G_r = \{1\}$$

of subgroups of G with G_i a normal subgroup of G_{i-1} (for $i = 1, \ldots, r$).

Definition 15.2 *A* normal series *for G is a chain*

$$G = G_0 \geq G_1 \geq G_2 \geq \cdots \geq G_r = \{1\},$$

of normal subgroups of G.

Note that every normal series is subnormal, but Example 15.4 below shows that the converse is not true.

Example 15.3 Let $G = \langle x \rangle$ be the cyclic group of order 6. Then

$$G \geq \langle x^2 \rangle \geq \{1\}$$

and

$$G \geq \langle x^3 \rangle \geq \{1\}$$

are both normal series for G since any subgroup of an abelian group is normal.

Example 15.4 Let G be the alternating group $A(4)$ and V be the subset $\{1, (1\ 2)(3\ 4), (1\ 3)(2\ 4), (1\ 4)(2\ 3)\}$. It may easily be checked that V is a subgroup by completing the 4×4 multiplication table for these elements. In fact V is a normal subgroup since any conjugate of an element of cycle type $(i\ j)(k\ l)$ will have the same cycle type by Proposition 9.20. However, the three non-identity elements of V are the only elements in $A(4)$ of this cycle type, so V is a normal subgroup of G. Since V is abelian, the subgroup $H = \{1, (1\ 2)(3\ 4)\}$ is a normal subgroup of V and so

$$G \geq V \geq H \geq \{1\}$$

is a subnormal series for G. This is not a normal series because H is not a normal subgroup of G, since, for example,

$$(1\ 2\ 3)(1\ 2)(3\ 4)(1\ 2\ 3)^{-1} = (2\ 3)(1\ 4)$$

is not an element of H.

Definition 15.5 *Let*

$$G = G_0 \geq G_1 \geq G_2 \geq \cdots \geq G_r = \{1\} \qquad (A)$$

and

$$G = H_0 \geq H_1 \geq H_2 \geq \cdots \geq H_s = \{1\} \qquad (B)$$

both be subnormal series of G. Then (B) is said to be a subnormal refinement *of (A) if each group which appears in (A) also occurs in (B). Similarly, if (A) and (B) are both normal series for G, (B) is a* normal refinement *of (A) if each group which appears in (A) also occurs in (B).*

Definition 15.6 *The series (A) and (B) are* isomorphic *if there is a bijection between the sets*

$$\{G_0/G_1, G_1/G_2, \ldots, G_{r-1}/G_r\} \text{ and } \{H_0/H_1, H_1/H_2, \ldots, H_{s-1}/H_s\}$$

of quotient groups such that groups which correspond under the bijection are isomorphic. Thus, in particular, $r = s$. Note that this definition applies to the case when (A) and (B) are both subnormal series and also to the case when (A) and (B) are normal series.

Example 15.7 Let $G = \langle x \rangle$ be a cyclic group of order 30, so that every subgroup of G is normal. Then $G > \{1\}$ is a normal series for G. Let $H = \langle x^2 \rangle$, so that H has order 15, then $G > H > \{1\}$ is a refinement of the normal series. This series may itself be refined by adding the normal subgroup $K = \langle x^6 \rangle$ to give

$$G > H > K > \{1\}.$$

This normal series cannot be further refined without repeating terms.

Example 15.8 In the case of a cyclic group $G = \langle x \rangle$ of order 6, the two series

$$G \geq \langle x^2 \rangle \geq \{1\}$$

and

$$G \geq \langle x^3 \rangle \geq \{1\}$$

which we have already considered are isomorphic, because

$$G/\langle x^2 \rangle \cong \langle x^3 \rangle \text{ and } G/\langle x^3 \rangle \cong \langle x^2 \rangle.$$

The first step in the proof of the Jordan–Hölder Theorem is another isomorphism theorem, sometimes known as the Zassenhaus lemma. This result will only be needed once, for the proof of Theorem 15.10 below. It is possible to defer reading the details of the proof of this result at this stage.

Proposition 15.9 *Let H, H_1 and K, K_1 be subgroups of a group G with H_1 a normal subgroup of H and K_1 a normal subgroup of K. Then*

$$\frac{H_1(H \cap K)}{H_1(H \cap K_1)} \cong \frac{K_1(H \cap K)}{K_1(H_1 \cap K)}.$$

Proof This is a somewhat technical result and includes the fact that the quotient groups displayed in the statement are well-defined. Since H_1 is a normal subgroup of H, it follows that $H \cap K$ (and H_1) both normalize H_1. Hence H_1 is a normal subgroup of $\langle H_1,\ H \cap K \rangle$. Thus, by Proposition 7.8,

$$\langle H_1, H \cap K \rangle = H_1(H \cap K) = (H \cap K)H_1.$$

By the First Isomorphism Theorem, there is a map

$$\vartheta : (H \cap K)H_1/H_1 \to (H \cap K)/(H \cap K \cap H_1) = (H \cap K)/(H_1 \cap K),$$

given by

$$\vartheta(hH_1) = h(H_1 \cap K) \text{ for all } h \in H \cap K.$$

Since $H \cap K_1$ and $H_1 \cap K$ are normal subgroups of $H \cap K$ by Proposition 7.8, an application of Proposition 7.9 shows that $(H_1 \cap K)(H \cap K_1)$ is a normal subgroup of $H \cap K$. The proof of the Second Isomorphism Theorem then shows that there is a homomorphism

$$\phi : (H \cap K)/(H_1 \cap K) \to (H \cap K)/(H \cap K_1)(H_1 \cap K)$$

defined by

$$\phi(h(H_1 \cap K)) = h(H \cap K_1)(H_1 \cap K) \text{ for all } h \in H \cap K,$$

with the kernel of ϕ being $(H \cap K_1)(H_1 \cap K)/(H_1 \cap K)$. Consider the composite map

$$\phi \circ \vartheta : (H \cap K)H_1/H_1 \to (H \cap K)/(H \cap K_1)(H_1 \cap K)$$

The kernel of this map is the set of cosets $hH_1 \in (H \cap K)H_1/H_1$ such that $\phi \circ \vartheta(hH_1)$ is the identity element of the group $(H \cap K)/(H \cap K_1)(H_1 \cap K)$. Thus $\ker \phi \circ \vartheta$ is given by

$$\begin{aligned} & \{hH_1 \in (H \cap K)H_1/H_1 : \phi(h(H_1 \cap K)) = (H \cap K_1)(H_1 \cap K)\} \\ = \; & \{hH_1 \in (H \cap K)H_1/H_1 : h \in (H \cap K_1)(H_1 \cap K)\} \\ = \; & \{hH_1 \in (H \cap K)H_1/H_1 : hH_1 \in (H \cap K_1)H_1\} \\ = \; & (H \cap K_1)H_1/H_1. \end{aligned}$$

Since ϕ and ϑ are both surjective, so is $\phi \circ \vartheta$ and we can apply the Homomorphism Theorem to deduce that

$$\frac{(H \cap K)H_1/H_1}{(H \cap K_1)H_1/H_1} \cong (H \cap K)/(H \cap K_1)(H_1 \cap K).$$

Another application of the Second Isomorphism Theorem gives that

$$\frac{(H \cap K)H_1/H_1}{(H \cap K_1)H_1/H_1} \cong \frac{(H \cap K)H_1}{(H \cap K_1)H_1}$$

so that

$$\frac{(H \cap K)H_1}{(H \cap K_1)H_1} \cong (H \cap K)/(H \cap K_1)(H_1 \cap K).$$

In particular, we have shown that the quotient group

$$\frac{(H \cap K)H_1}{(H \cap K_1)H_1}$$

is well-defined. Now repeating the proof with H and K interchanged, we see that

$$(H \cap K)/(H \cap K_1)(H_1 \cap K) \cong \frac{(H \cap K)K_1}{(H_1 \cap K)K_1},$$

and this gives the required result. □

We now come to the second step in the proof of the Jordan–Hölder Theorem.

Theorem 15.10 (Schreier's Refinement Theorem). *Any two subnormal series of a group G have subnormal refinements which are isomorphic. Similarly, any two normal series of a group G have isomorphic normal refinements.*

Proof Suppose we are given two subnormal series

$$G = G_0 \geq G_1 \geq G_2 \geq \cdots \geq G_r = \{1\} \tag{A}$$

and

$$G = H_0 \geq H_1 \geq H_2 \geq \cdots \geq H_s = \{1\} \tag{B}$$

for G. We shall refine the series (A) to a series (A)$'$ and show that this is isomorphic to a refinement (B)$'$ of (B). To do this, define $G_{i,j}$ to be $(G_i \cap H_j)G_{i+1}$, for j in $\{0, \ldots, s\}$ and for i in $\{0, \ldots, r-1\}$. Thus

$$G_{i,0} = (G_i \cap G)G_{i+1} = G_i$$

and

$$G_{i,s} = (G_i \cap \{1\})G_{i+1} = G_{i+1},$$

so that for each i, we obtain a refinement

$$G_{i,0} \geq G_{i,1} \geq \cdots \geq G_{i,s}$$

of the terms between G_i and G_{i+1}. Extend this to a refinement (A)$'$ of the series (A) by repeating this for each i. However, since $G_{i,s} = G_{i+1} = G_{i+1,0}$, we can omit the terms $G_{i,s}$ except when $i = r-1$. Thus there are $rs+1$ groups in the series (A)$'$.

Similarly for j in $\{0, \ldots, s-1\}$, we obtain a refinement

$$H_{0,j} \geq H_{1,j} \geq \cdots \geq H_{r,j}$$

of the terms between H_j and H_{j+1} by defining $H_{i,j}$ to be $(G_i \cap H_j)H_{j+1}$ (for $0 \leq i \leq r$). After deleting the terms $H_{r,j}$ except when j is $s-1$, this gives a refinement (B)$'$ of (B). The fact that the series (A)$'$ and (B)$'$ are isomorphic now follows by Proposition 15.9 since

$$G_{i,j}/G_{i,j+1} \cong H_{i,j}/H_{i+1,j}$$

for $0 \leq i \leq r-1$ and $0 \leq j \leq s-1$.

Finally, notice that if the original series were a normal series for G, then each $G_{i,j}$ and each $H_{i,j}$ would be a normal subgroup of G, so the refined series would also be a normal series. □

Example 15.11 Let G be the infinite cyclic group $\langle x \rangle$. Since G is abelian, each subnormal series is a normal series. For any positive integer d, let $G(d)$ denote the subgroup generated by x^d. Consider the (sub)normal series

$$G \geq G(2) \geq G(4) \geq \{1\},$$

and

$$G \geq G(5) \geq G(10) \geq \{1\}.$$

These have isomorphic refinements

$$G \geq G(2) \geq G(4) \geq G(20) \geq G(40) \geq \{1\}$$

and

$$G \geq G(5) \geq G(10) \geq G(20) \geq G(40) \geq \{1\},$$

in which the relevant quotient groups are $\{C_2, C_2, C_5, C_2, C_\infty\}$ and $\{C_5, C_2, C_2, C_2, C_\infty\}$.

Definition 15.12 *A* composition series *for a group G is a subnormal series without repetitions which can be refined only by repeating terms.*

Definition 15.13 *A* chief series *for G is a normal series without repetitions which can be refined (by a normal series) only by repeating terms.*

Remark 1 The infinite cyclic group $G = \langle x \rangle$ has neither a composition series nor a chief series. To see this suppose that

$$G = G_0 > G_1 > G_2 > \cdots > G_r = \{1\}$$

is a composition series for G. Since G is abelian, this would also be a chief series. Since each subgroup of a cyclic group is cyclic (Proposition 4.13), the group G_{r-1} is cyclic generated by x^s for some s. This means that G_{r-1} is actually an infinite cyclic group, so that $\langle x^{2s} \rangle$ is a proper subgroup of G_{r-1}. We would then obtain a non-trivial refinement

$$G = G_0 > G_1 > G_2 > \cdots > G_{r-1} > \langle x^{2s} \rangle > \{1\}$$

of the given series. Thus the infinite cyclic group does not have a composition series. However, every *finite* group has both composition and chief series, since the process of refining the series $G \geq \{1\}$ must terminate in a finite number of steps.

Remark 2 Suppose that we are given a series

$$G = G_0 \geq G_1 \geq G_2 \geq \cdots \geq G_r = \{1\}$$

for G which may be either normal or subnormal. If we know that for some value of i, the index of G_{i+1} in G_i is a prime integer p, then the series cannot be refined between these terms. This is because any subgroup H with $G_i \geq H \geq G_{i+1}$ would give rise to a subgroup H/G_{i+1} of the group G_i/G_{i+1}, by the Correspondence Theorem 7.14. Since this latter group has p elements, this is only possible if H is either G_i or G_{i+1}.

Example 15.14 The normal series $G \geq \langle x^2 \rangle \geq \{1\}$ for the cyclic group with six elements is a chief series. Since $\langle x^2 \rangle$ is of order three, the only subgroups H satisfying $\langle x^2 \rangle \geq H \geq \{1\}$ are $\langle x^2 \rangle$ and $\{1\}$. Similarly, any subgroup H satisfying $G \geq H \geq \langle x^2 \rangle$ has order divisible by 3 and dividing 6, so H is either G or $\langle x^2 \rangle$.

Example 15.15 In $A(4)$, let V be the subgroup

$$\{1, (1\ 2)(3\ 4), (1\ 3)(2\ 4), (1\ 4)(2\ 3)\}.$$

Then the subnormal series

$$A(4) \geq V \geq \{1, (1\ 2)(3\ 4)\} \geq \{1\}$$

is a composition series because the indices

$$|A(4) : V|,\ |V : \{1, (1\ 2)(3\ 4)\}|, \text{ and } |\{1, (1\ 2)(3\ 4)\} : \{1\}|$$

are all prime integers. However, this is not a chief series since, as we have already seen, $\{1, (1\ 2)(3\ 4)\}$ is not a normal subgroup of $A(4)$. In fact, it can be checked that none of the three subgroup of V of order 2 is normal in $A(4)$, so that there are no proper normal subgroups of $A(4)$ between V and $\{1\}$. It follows that a chief series for $A(4)$ is $A(4) \geq V \geq \{1\}$.

We now come to the main result of this section.

Theorem 15.16 (The Jordan–Hölder Theorem). *If a group has a composition series then any two composition series are isomorphic. A similar result holds for chief series.*

Proof The proof is an easy application of the Schreier Refinement Theorem. Suppose that

$$G = G_0 > G_1 > G_2 > \cdots > G_r = \{1\} \tag{A}$$

and

$$G = H_0 > H_1 > H_2 > \cdots > H_s = \{1\} \tag{B}$$

are both composition series for G. By Theorem 15.10, these have isomorphic refinements. However, by definition, they cannot be refined without repeating terms. It follows that the series must be isomorphic in the first place. The proof for chief series is similar. □

Example 15.17 We have already seen that $G \geq \langle x^2 \rangle \geq \{1\}$ is a chief series for G, the cyclic group of order 6. Another chief series, which is isomorphic to this, is $G \geq \langle x^3 \rangle \geq \{1\}$.

Summary for Chapter 15

In this chapter, we consider *composition series* for a group G. These are descending chains of distinct subgroups from G to $\{1\}$ in which each term is a normal subgroup of the group immediately above it. In addition, further terms cannot be inserted without repeating existing terms. A given group may have several distinct composition series, but the Jordan–Hölder Theorem tells us that in any two such series, the sets of associated quotient groups are isomorphic.

The first result in this direction was published by Jordan in 1870; however, Jordan had no concept of a quotient group available to him. Then in 1889, Hölder proved that the groups in a composition series are isomorphic. The formal idea of refinement was introduced by Schreier in 1926, and the isomorphism result, Proposition 15.9, was proved by Zassenhaus in 1934.

Exercises 15

1. Find composition series and chief series for

(a) the symmetric group $S(4)$;
(b) the quaternion group of order 8 with presentation

$$\langle a, b : a^4 = 1, a^2 = b^2, ba = ab^{-1} \rangle;$$

(c) the dihedral group $D(6)$.

2. Find composition series and chief series for the group $A(4) \times A(4)$. [Hint: use the fact, proved in Exercises 11, that if N_1 and N_2 are normal subgroups of the groups G_1 and G_2, respectively, then $N_1 \times N_2$ is a normal subgroup of $G_1 \times G_2$.]

3. Give examples of:

(a) a group with a normal series with an infinite number of terms;
(b) a group with a composition series which is not a chief series; and
(c) a group with two different chief series.

16
Composition factors and chief factors

In this chapter we shall study the quotient groups that occur in composition series and chief series. As usual, there are two related definitions.

Definition 16.1 *Suppose that*

$$G = G_0 > G_1 > \cdots > G_{r-1} > G_r = \{1\}$$

is a composition series for the group G. The quotient groups

$$G_0/G_1, G_1/G_2, \ldots, G_{r-1}/G_r$$

are the composition factors *of G.*

Definition 16.2 *If*

$$G = G_0 > G_1 > \cdots > G_{r-1} > G_r = \{1\}$$

is a chief series for the group G, the quotient groups

$$G_0/G_1, G_1/G_2, \ldots, G_{r-1}/G_r$$

are the chief factors *of G.*

Remark It follows by the Jordan–Hölder Theorem that the set of composition factors of a given group G is independent of the composition series and this set is therefore an invariant of the group G. A similar remark applies to the chief factors of a group.

We now introduce one of the most important concepts in the subject.

Definition 16.3 *A group G is* simple *if the only normal subgroups of G are $\{1\}$ and G.*

Proposition 16.4 *Any composition factor of a group is a simple group.*

Proof The proof is an easy application of the Correspondence Theorem 7.14. Suppose that G_i/G_{i+i} is a non-simple composition factor of a group G. Thus there is a proper normal subgroup of G_i/G_{i+1}. By the Correspondence Theorem, this is a subgroup N/G_{i+1} giving a chain $G_i > N > G_{i+1}$ with N a normal subgroup of G_i. This contradicts the fact that a composition series cannot be refined, so we conclude that the composition factors must be simple groups. □

Remark Since we may regard the group as being built up from its composition factors, the simple groups are the 'atomic' objects from which other groups are constructed. The complete list of finite simple groups is known: this is the Classification Theorem for finite simple groups. We shall have more to say about this in Chapter 25. There is one situation in which the simple groups are easy to classify.

Proposition 16.5 *A simple abelian group is cyclic of prime order. In particular, a composition factor of a finite abelian group is cyclic of prime order.*

Proof Suppose that G is a simple abelian group and x is any non-identity element of G. The cyclic subgroup $\langle x\rangle$ is normal in G and is not $\{1\}$, so it is G. Now consider the subgroup $\langle x^2\rangle$ of the cyclic group G. If this is $\{1\}, G$ is cyclic of order 2. Otherwise, the simplicity of G forces $\langle x^2\rangle$ to equal G. It follows that x is in $\langle x^2\rangle$ and so is of the form x^{2n} for some n. Thus x has finite order and hence G is finite. Finally, let p be any prime divisor of the order of the finite cyclic group G. It follows by Proposition 4.14 that G has a subgroup H, say, of order p. The fact that G is simple shows that $G = H$ and so G has order p. The consequence for composition factors then follows by Proposition 16.4. □

Remark There is a sense in which the unique factorisation theorem for integers can be considered as a special case of the Jordan–Hölder Theorem. For any positive integer n, consider a composition series for the cyclic group G with n elements

$$G = G_0 > G_1 > \cdots > G_{r-1} > G_r = \{1\}.$$

Since G is abelian, its composition factors are all cyclic of prime order by Proposition 16.5, and so $n = |G|$ can be written as a product of primes. Conversely, given any decomposition of $|G|$ as a product of primes, we can write a composition series for G with a factor for each prime which occurs in the given decomposition (including multiplicities). Since these composition factors are independent of the particular composition series, this decomposition of n into a product of primes is unique.

The next objective will be to explain the structure of the chief factors of a group. As we have already seen, these need not be simple groups. A preliminary idea is needed.

Definition 16.6 *A subgroup H of a group G is* characteristic *if for each automorphism ϑ of $G, \vartheta(H) = H$.*

Remark Notice that for any fixed element $g \in G$, the map ϑ_g defined by $\vartheta_g(x) = gxg^{-1}$ for all $x \in G$ is an automorphism of G. It follows that any characteristic subgroup is necessarily a normal subgroup.

Example 16.7 Suppose that H is a finite subgroup of order k of a group G and suppose that H is the only subgroup of G with k elements. Then since an automorphism is a bijection, $|\vartheta(H)| = |H|$ for all automorphisms ϑ of G. It follows that $\vartheta(H) = H$ for all $\vartheta \in \mathrm{Aut}(G)$. Thus H is a characteristic subgroup of G in this case. Examples of this situation occur when G is a finite group which has a unique Sylow p-subgroup P for some p dividing $|G|$, for example in $S(3)$ when $p = 3$ and in $A(4)$ when $p = 2$.

Proposition 16.8 *Let N be a normal subgroup of a group G and let K be a characteristic subgroup of N. Then K is a normal subgroup of G.*

Proof For any g in G, conjugation of N by g is an automorphism of N, and so K is invariant under the maps $\vartheta_g(x) = gxg^{-1}$, for all $g \in G$. □

Definition 16.9 *A group G is* characteristically simple *if the only characteristic subgroups of G are $\{1\}$ and G itself.*

We can now give the basic property of chief factors.

Proposition 16.10 *Every chief factor of a group G is characteristically simple.*

Proof Let G_i/G_{i+1} be a chief factor of G, and suppose that K/G_{i+1} is a characteristic subgroup of G_i/G_{i+1}. By Proposition 16.8, K/G_{i+1} is a normal subgroup of G/G_{i+1}, and so K is a normal subgroup of G. Since the chief series cannot be refined between G_i and G_{i+1}, we conclude that K is equal to either G_i or G_{i+1}, and so the given chief factor is characteristically simple. □

It turns out that characteristically simple groups are closely related to simple groups.

Proposition 16.11 *A finite characteristically simple group is a direct product of isomorphic simple groups. In particular, a chief factor of a finite group is a direct product of isomorphic simple groups.*

Proof Let G be a characteristically simple group and N be a minimal normal subgroup of G, so that there is no normal subgroup N_0 of G with $N > N_0 > \{1\}$. Let M be a subgroup of G of largest order for which M is an internal direct product of subgroups $N_1, N_2, \ldots, N_r$ of G with the property that $N_i = \vartheta_i(N)$ with $\vartheta_1, \vartheta_2, \ldots, \vartheta_r$ automorphisms of G. Since each $\vartheta_i(N)$ is a normal subgroup of G, M is a normal subgroup of G.

We first show that for all automorphisms ϕ of G, $\phi(N)$ is contained in M. Otherwise, we can find $\phi \in \mathrm{Aut}(G)$ such that $K = \phi(N)$ is not contained in M. Then $M \cap K$ is an intersection of two normal subgroups of G, and so is itself a normal subgroup of G. Since N is a minimal normal subgroup of G, so is $K = \phi(N)$. However, $M \cap K$ is not equal to K, and so the minimality of K forces this intersection to be $\{1\}$. Proposition 7.10 then shows that $\langle M, K \rangle = MK$, and it then follows by Proposition 13.5 that $\langle M, K \rangle$ is an internal direct product $M \times K$, contrary to the definition of M.

We may therefore suppose that for any $\phi \in \mathrm{Aut}(G)$, the subgroup $\phi(N)$ is contained in M. Since by definition M is generated by groups isomorphic to N, we have that $M = \langle \phi(N) : \phi \in \mathrm{Aut}(G) \rangle$. Thus M is a characteristic subgroup of G and is not $\{1\}$, so M is G. We now know that G is the direct product of groups $N_1 \times N_2 \times \cdots \times N_r$ with N_i isomorphic to the normal subgroup N. We can therefore deduce that N is simple, because if N_0 were a proper normal subgroup of N, and $N_1 = \vartheta(N)$, then $\vartheta(N_0)$ would be a normal subgroup of N_1. Since G is an internal direct product of $N_1, \ldots, N_r$, it follows by Corollary 13.6 that $\vartheta(N_0)$ is a normal subgroup of G, contrary to minimality of N. □

Corollary 16.12 *A finite abelian chief factor of a group is an elementary abelian p-group for some prime p.*

Proof The proof follows directly using Propositions 16.11 and 16.5. □

To conclude this section, we demonstrate the existence of some non-abelian simple groups by showing that the alternating groups $A(n)$ are simple for $n \geq 5$. This proof will be by induction on n, and the next result provides the anchor step for this induction.

Theorem 16.13 *The alternating group $A(5)$ is a non-abelian simple group.*

Proof It is clear that $G = A(5)$ is non-abelian, since the elements $(1\ 2\ 3)$ and $(1\ 2)(3\ 4)$ do not commute. We show that it is simple by considering the partition of the 60 elements of G into conjugacy classes. It will transpire that there are five classes C_1, C_2, C_3, C_4 and C_5, with C_1 being $\{1\}$. If N were a normal subgroup, then for each element x of N and each g in G, the conjugate gxg^{-1} would also be in N. Thus N would be a union of some of the conjugacy classes of G, and also N would necessarily contain C_1. Once the list of conjugacy classes of G is known, it will be clear that the number of elements in each proper subset of these which include C_1 is never a divisor of 60. It will follow that no union of conjugacy classes can form a proper subgroup of G, so that G is a simple group.

To investigate the conjugacy classes of G, first consider the class of the element $(1\ 2)(3\ 4)$. The number of elements which are products of two disjoint transpositions is calculated as follows: there are 5 choices for the 'first' entry a in $(a\, b)(c\, d)$, 4 choices for the next, 3 for the next and 2 choices for the last. However, interchanging a and b gives the same permutation, as does interchanging c and d, and interchanging the transpositions $(a\ b)$ and $(c\ d)$ also gives the same permutation. The total number of elements of cycle type 2^2 is therefore $(5 \times 4 \times 3 \times 2)/(2 \times 2 \times 2) = 15$. By Proposition 9.20, any conjugate of $\rho = (1\ 2)(3\ 4)$ has cycle type 2^2 (so that each element is a product of two disjoint transpositions), hence the conjugacy class C_2 of ρ has at most 15 elements. To determine the exact size of C_2, we show that the centralizer $C_G(\rho)$ has 4 elements, so that C_2 has $60/4 = 15$ elements and hence consists of all elements of cycle type 2^2. If

$$\pi = \begin{pmatrix} 1 & 2 & 3 & 4 & 5 \\ a & b & c & d & e \end{pmatrix}$$

is in the centraliser of $\rho = (1\ 2)(3\ 4)$, then $\phi = \pi\rho\pi^{-1} = (a\ b)(c\ d)$. It follows that none of a, b, c or d can equal 5, so $e = 5$ and π fixes 5. Thus either $(a\ b) = (1\ 2)$ (in which case $(c\ d) = (3\ 4)$) or $(a\ b) = (3\ 4)$ (in which case $(c\ d) = (1\ 2)$). This gives 8 possibilities for elements in $C_G(\rho)$:

$$1,\ (1\ 2),\ (3\ 4),\ (1\ 2)(3\ 4),$$

$$(1\ 3)(2\ 4),\ (1\ 4)(2\ 3),\ (1\ 3\ 2\ 4) \text{ and } (1\ 4\ 2\ 3).$$

Of these only the elements

$$1,\ (1\ 2)(3\ 4),\ (1\ 3)(2\ 4) \ \text{ and } \ (1\ 4)(2\ 3)$$

are even permutations, so the centraliser of ρ in $A(5)$ does indeed have 4 elements. We have therefore proved that there are 15 elements in the conjugacy class of $(1\ 2)(3\ 4)$.

Next, we calculate the number of conjugates of a 3-cycle in G. Let ρ be the 3-cycle $(1\,2\,3)$. The total number of 3-cycles in G is $(5 \times 4 \times 3)/3 = 20$. Using the element π of the previous paragraph, we see that $\pi\rho\pi^{-1} = (a\,b\,c)$. Thus the centraliser of ρ consists of the permutations

$$1,\ (1\,2\,3),\ (1\,3\,2),\ (4\,5),\ (1\,2\,3)(4\,5) \text{ and } (1\,3\,2)(4\,5).$$

Of these six, the first three are even and the next three are odd, so the centraliser of ρ in G consists of the three permutations $1, (1\,2\,3)$ and $(1\,3\,2)$. It follows that ρ has $60/3 = 20$ conjugates and so the conjugacy class C_3 of a 3-cycle contains all 20 3-cycles.

We now turn to the conjugates of a 5-cycle $\rho = (1\ 2\ 3\ 4\ 5)$. The total number of 5-cycles is $(5 \times 4 \times 3 \times 2 \times 1)/5 = 24$. If π is the permutation

$$\begin{pmatrix} 1 & 2 & 3 & 4 & 5 \\ a & b & c & d & e \end{pmatrix},$$

then $\pi\rho\pi^{-1} = (a\,b\,c\,d\,e)$. The elements in the centraliser of ρ are therefore determined by specifying the value of a, which determines the values of b, c, d and e. This gives five permutations:

$$1,\ (1\,2\,3\,4\,5),\ (1\,3\,5\,2\,4),\ (1\,4\,2\,5\,3) \text{ and } (1\,5\,4\,3\,2).$$

Thus ρ has $60/5 = 12$ conjugates. This means that half the 24 5-cycles are conjugate to ρ. If we choose one of the 12 5-cycles not conjugate to ρ, the same argument shows that this element will have 12 conjugates. Since conjugacy classes partition the group, it follows that the 5-cycles divide into two classes C_4 and C_5, each with 12 elements.

We have therefore found five conjugacy classes of G which contain a total of $1 + 15 + 20 + 12 + 12 = 60$ elements, so this is the complete list of conjugacy classes of G. It can be seen that any subset of this list of conjugacy classes containing C_1 can only contain a total of d elements, where d divides 60, if d is 1 or 60. It follows that $A(5)$ has no proper normal subgroups and so it is a non-abelian simple group. □

The next objective is to show that the alternating group $A(n)$ is simple for $n \geq 5$, but we need a preliminary result.

Proposition 16.14 *The alternating group $A(n)$ is generated by its 3-cycles.*

Proof Let π be an element of $A(n)$. We know that, considered as an element of $S(n), \pi$ can be written as a product of an even number of transpositions. The result will therefore be proved once we show that any product of two distinct transpositions can be written as a product of 3-cycles. We therefore consider the product $(i\ j)(r\ s)$. If the four integers i, j, r, s are distinct, we note that

$$(i\ j)(r\ s) = (i\ r\ j)(i\ r\ s),$$

as required. Otherwise, we may suppose without loss of generality that $i = r$, in which case $(i\ j)(i\ s) = (i\ s\ j)$, as required. □

Corollary 16.15 *The only normal subgroup of $A(n)$ which contains a 3-cycle is $A(n)$ itself.*

Proof Let N be a normal subgroup of $A(n)$ which contains a 3-cycle π, so that $n \geq 3$. If $n = 3$ or 4, we see by inspection that N is the group $A(n)$. We therefore suppose that $n \geq 5$. Since N is a normal subgroup, N contains the conjugacy class $\mathcal{C}$ of π in $A(n)$. The conjugacy class of π in $S(n)$ contains all 3-cycles by Proposition 9.20. We shall show that the conjugacy class of π in $A(n)$ contains all 3-cycles, so that N is equal to $A(n)$ by Proposition 16.14. It is possible to do this by showing directly that any two 3-cycles are conjugate in $A(n)$, but several cases need to be distinguished, so we give a more elegant proof. Since $n \geq 5$, given any 3-cycle π (for example $(1\ 2\ 3)$) there is an odd permutation (for example $(4\ 5)$) which centralises the 3-cycle. It follows that

$$C_{S(n)}(\pi) > C_{A(n)}(\pi).$$

However, by Proposition 10.25,

$$C_{A(n)}(\pi) = C_{S(n)}(\pi) \cap A(n).$$

Since there is an odd permutation in $C_{S(n)}(\pi)$, it follows that

$$\begin{aligned} |C_{S(n)}(\pi) : C_{A(n)}(\pi)| &= |C_{S(n)}(\pi) : C_{S(n)}(\pi) \cap A(n)|. \\ &= |A(n)C_{S(n)}(\pi) : A(n)| \quad \text{by Proposition 5.18} \\ &= |S(n) : A(n)| \\ &= 2. \end{aligned}$$

It now follows that the number $|A(n) : C_{A(n)}(\pi)|$ of conjugates of π in $A(n)$ is equal to the number $|S(n) : C_{S(n)}(\pi)|$ of conjugates of π in $S(n)$, so both of these classes coincide with the set of all 3-cycles, as required. □

Theorem 16.16 *The group $A(n)$ is simple for $n \geq 5$.*

Proof The proof is by induction on n, the case $n = 5$ being Theorem 16.13. Suppose that $n > 5$ and let N be a normal subgroup of $A(n)$. We first show that N must contain an element which fixes some integer $i \in \{1, \ldots, n\}$. To do this, suppose that every non-identity element of N has no fixed points, and let π be such an element. Thus, denoting $\pi(1)$ by a, we have that $a \neq 1$. Now we consider the two-row form of the permutation π and choose a column not containing 1 or a. Thus we choose integers b and c different from both 1 and a with $\pi(b) = c$. Also, since π has no fixed points, $b \neq c$. Since $n \geq 6$, we can choose integers d and e distinct from $1, a, b$ and c, and so $\rho = (1\ a)(b\ c\ d\ e)$ is an element of $A(n)$. Since N is a normal subgroup of $A(n)$, $\rho\pi\rho^{-1}$ is an element of N, and $\rho\pi\rho^{-1}(a) = 1$, whereas $\rho\pi\rho^{-1}(c) = d$. Now $(\rho\pi\rho^{-1})\pi$ is an element of N and

$$(\rho\pi\rho^{-1})\pi(1) = \rho\pi\rho^{-1}(a) = 1.$$

Thus $(\rho\pi\rho^{-1})\pi$ fixes 1 and is not the identity since

$$(\rho\pi\rho^{-1})\pi(b) = \rho\pi\rho^{-1}(c) = d.$$

Then $\rho\pi\rho^{-1}\pi$ is a non-identity element of N which fixes 1. This contradiction shows that there is an element of N which has at least one fixed point. We therefore suppose that N contains an element which fixes an integer i.

Now let A_i be the subgroup of $A(n)$ consisting of all those permutations which fix i, so that A_i is isomorphic to $A(n-1)$. Then $N_i = N \cap A_i$ is a normal subgroup of A_i. By induction A_i is simple, and so we deduce that N_i is either $\{1\}$ or A_i. Since N contains a non-identity element which fixes i, N_i contains more than one element, so $N_i = A_i$. It follows that A_i is contained in N. Since $n > 5$, there is a 3-cycle π in A_i, and hence in N, so that $N = A(n)$ in this case by Corollary 16.15. Thus we conclude that $A(n)$ is simple. □

Remark We shall see other examples of finite simple groups in later chapters. We shall also see in Chapter 18 that 60 is the smallest possible order of a non-abelian finite simple group.

Summary for Chapter 16

The Jordan–Hölder Theorem implies that the set of composition factors (the groups G_0/G_1, $G_1/G_2, \ldots, G_{r-1}/G_r$) in a composition series for G is independent of the particular series, and so depends only on the group G. Since the composition factors are simple groups, the finite simple groups are the building blocks from which finite groups are constructed. The complete list of finite simple groups is known, and this will be discussed later.

It is also shown that any chief factor of a group is a direct product of isomorphic simple groups. The simple abelian groups are cyclic of prime order, and the easiest example of a non-abelian simple group is the alternating group $A(5)$. The groups in which each composition factor is abelian can be expected to be of interest. These groups are *soluble groups* and will be studied in the next chapter.

Jordan (1870) proved that $A(n)$ is generated by 3-cycles and also that $A(n)$ is the only non-trivial normal subgroup of $S(n)$ for $n \geq 5$.

Exercises 16

1. Give an example of two non-isomorphic groups with isomorphic chief series.

2. Find all the characteristic subgroups of the symmetric group $S(3)$, the dihedral group $D(4)$ and the quaternion group of order 8.

3. Give an example of an integer n for which there are two groups of order n with composition series of different lengths.

4. Let H be a characteristic subgroup of G and K be a characteristic subgroup of H. Show that K is a characteristic subgroup of G.

17
Soluble groups

In this chapter, we shall introduce the concept of a *soluble* group and establish some of the basic properties of such groups.

Definition 17.1 *A group G is said to be* soluble *if G has a normal series*

$$G = G_0 \geq G_1 \geq G_2 \geq \cdots \geq G_r = \{1\}$$

in which the quotient groups G_0/G_1, G_1/G_2, ..., G_{r-1}/G_r are abelian groups.

Example 17.2 It is clear from the definition that every abelian group is soluble, so that the concept of solubility may be regarded as a generalisation of being abelian. Although there is no complete classification theorem for finite soluble groups, there are several basic properties which are exactly analogous to properties of abelian groups. For example, we shall prove in Proposition 17.5 below that subgroups and quotient groups of soluble groups are soluble.

Example 17.3 We saw in Example 15.4 that the series

$$A(4) \geq V \geq \{1\}$$

is a normal series (in fact a chief series) for $A(4)$. The quotient group $A(4)/V$ is of order 3 and is therefore abelian. Since V is abelian, the quotient group $V/\{1\} = V$ is also abelian. Thus $A(4)$ is soluble.

Example 17.4 The group $S(5)$ has $A(5)$ as a normal subgroup of index 2, and by Theorem 16.13, $A(5)$ is a non-abelian simple group. Thus

$$S(5) \geq A(5) \geq \{1\}$$

is a chief series for $S(5)$. It follows that $A(5)$ is a non-abelian chief factor of $S(5)$. If $S(5)$ were soluble, there would be a normal series

$$S(5) = G_0 \geq G_1 \geq G_2 \geq \cdots \geq G_r = \{1\}$$

for $S(5)$ with abelian factors. Refining this to a chief series would give a chief series with abelian factors. However the series

$$S(5) \geq A(5) \geq \{1\}$$

cannot be refined to a series with abelian factors. It follows using the Jordan–Hölder Theorem that the group $S(5)$ is not soluble.

Proposition 17.5 *Let G be a soluble group. Then*

(i) *for any subgroup H of G, H is a soluble group, and*
(ii) *for any normal subgroup N of G, the quotient group G/N is soluble.*

Proof (i) Since G is soluble, there is a normal series

$$G = G_0 \geq G_1 \geq G_2 \geq \cdots \geq G_r = \{1\}$$

in which the quotient groups G_0/G_1, G_1/G_2, ..., G_{r-1}/G_r are abelian groups. Consider the series

$$H = H \cap G_0 \geq H \cap G_1 \geq H \cap G_2 \geq \cdots \geq H \cap G_r = \{1\}.$$

This is a normal series for H by Proposition 7.8. In order to show that H is soluble, it is sufficient therefore to show that for $i = 0, \ldots, r-1$, $H \cap G_i/H \cap G_{i+1}$ is abelian. To see this notice that

$$\begin{aligned} H \cap G_i/H \cap G_{i+1} &= H \cap G_i/((H \cap G_i) \cap G_{i+1}) \\ &\cong (H \cap G_i)G_{i+1}/G_{i+1} \quad \text{by Theorem 8.15} \\ &\leq G_i/G_{i+1}. \end{aligned}$$

Since subgroups of abelian groups are abelian, it follows that H has a normal series with abelian quotient groups, so that H is soluble.

(ii) To show that G/N is soluble, consider the series

$$G/N = G_0N/N \geq G_1N/N \geq G_2N/N \geq \cdots \geq G_rN/N = \{1\}.$$

Each G_iN/N is a normal subgroup of G/N by Proposition 7.14, so we only need to consider the successive quotient groups:

$$(G_iN/N)/(G_{i+1}N/N) \cong G_iN/G_{i+1}N \quad \text{by Theorem 8.16}$$

$$
\begin{aligned}
&= G_i(G_{i+1}N)/G_{i+1}N \\
&\cong G_i/(G_i \cap G_{i+1}N) \quad \text{by Theorem 8.15} \\
&\cong (G_i/G_{i+1})/((G_i \cap G_{i+1}N)/G_{i+1}),
\end{aligned}
$$

the last step being an application of Theorem 8.16, the Second Isomorphism Theorem. The result follows because quotient groups of abelian groups are abelian. □

Composition factors and chief factors of finite soluble groups can be described quite explicitly.

Proposition 17.6 *Let G be a finite soluble group. Then*

(i) *if G is simple, then G is cyclic of prime order;*
(ii) *any composition factor of G is cyclic of prime order;*
(iii) *a chief factor of G is an elementary abelian p-group for some prime p.*

Proof (i) By definition a soluble group has a normal series

$$G = G_0 \geq G_1 \geq G_2 \geq \cdots \geq G_r = \{1\}$$

with the quotient groups $G_0/G_1, G_1/G_2, \ldots, G_{r-1}/G_r$ abelian. If G is also simple, it follows that $r = 1$, and so G is abelian. Thus G is cyclic of prime order by Proposition 16.5.

(ii) Given that G is a soluble group, there is a normal series

$$G = G_0 \geq G_1 \geq G_2 \geq \cdots \geq G_r = \{1\}$$

with the quotients $G_0/G_1, G_1/G_2, \ldots, G_{r-1}/G_r$ being abelian groups. We can refine this series to a composition series for G and since subgroups and quotient groups of abelian groups are abelian, each composition factor of G will be abelian. The result will then follow by Propositions 16.4 and 16.5.

(iii) As in (ii), we can see that each chief factor of a soluble group is abelian. The result then follows by Corollary 16.12. □

The next objective is to give an alternative definition of solubility.

Definition 17.7 *Let x and y be elements of a group G. The* commutator *$[x, y]$ is the element $xyx^{-1}y^{-1}$.*

Definition 17.8 *The* commutator subgroup *or* derived group, *denoted by* $[G, G]$ *or* G', *is the subgroup generated by all commutators:*

$$[G, G] = G' = \langle [x, y] : x, y \in G \rangle.$$

Remark It is not always very easy to calculate the derived group from this definition. One common source of confusion lies in the fact that G' is defined to be the subgroup generated by the commutators (not just the set of commutators), thus, it is possible to have an element which is a product of commutators but which is not itself a commmutator. The next result gives a more practical method to determine G'.

Proposition 17.9 *For any group G the derived group G' is a characteristic subgroup of G and is the smallest normal subgroup of G with abelian quotient group, in the sense that if N is a normal subgroup of G with G/N abelian, then N contains G'.*

Proof Let x and y be any elements of G and let ϕ be any element of $\mathrm{Aut}(G)$. Then

$$\phi([x, y]) = \phi(xyx^{-1}y^{-1}) = \phi(x)\phi(y)\phi(x)^{-1}\phi(y)^{-1} = [\phi(x), \phi(y)],$$

so ϕ maps every generator of G' on to a generator of G'. This shows that G' is a characteristic subgroup of G.

Next, we show that G/G' is abelian: let $x, y \in G$, then since G' contains $[y^{-1}, x^{-1}] = y^{-1}x^{-1}yx$,

$$(xG')(yG') = xyG' = xy(y^{-1}x^{-1}yx)G' = yxG' = (yG')(xG'),$$

as required.

Finally, to show that G' is the smallest normal subgroup with abelian quotient, let N be a normal subgroup with G/N abelian. Thus, for all x and y in G,

$$yxN = (yN)(xN) = (xN)(yN) = xyN,$$

and so $xy(yx)^{-1} = [x, y]$ is an element of N. Since every generator of G' is in N, G' is contained in N, as claimed. □

Example 17.10 It is clear that a group G is abelian if and only if $G' = \{1\}$.

Example 17.11 Let G be the symmetric group $S(3)$. This group is non-abelian, so its derived group is non-trivial. The subgroup $A(3)$ has index 2 so the quotient $S(3)/A(3)$ is abelian, and G' is contained in $A(3)$. Since $A(3)$ is cyclic of order 3, the only subgroups of $A(3)$ are $A(3)$ and $\{1\}$. It follows that $G' = A(3)$.

Example 17.12 Let G be the alternating group $A(4)$ and V be the usual normal subgroup with the four elements $\{1, (1\,2)(3\,4), (1\,3)(2\,4), (1\,4)(2\,3)\}$. Then V is a minimal normal subgroup of G: the only normal subgroups of G contained in V are $\{1\}$ and V itself. Since $A(4)$ is non-abelian and $A(4)/V$ is abelian, G' is equal to V.

Remark Given a presentation for G, it is quite easy to obtain a presentation for the abelian quotient group $G/[G,G]$. This is done by adding to the presentation the relations saying that each pair of generators commute, thereby accounting for the fact that $G/[G,G]$ is abelian. Thus, for example, the dihedral group of order $2n$ has presentation

$$\langle x, y : x^n = 1 = y^2, xy = yx^{-1}\rangle.$$

Adding the relation $xy = yx$ to this presentation gives that $yx = yx^{-1}$ so that $x = x^{-1}$ and $x^2 = 1$. If n is odd, this is only possible if $x = 1$ and the commutator quotient group is therefore cyclic of order 2 (generated by the coset yG'). On the other hand, if n is even, we see that the commutator quotient group is non-cyclic of order 4 generated by the cosets of xG' and yG'.

Definition 17.13 *Define the* derived series *of G iteratively as follows:*

$$G^{(0)} = G;\; G^{(1)} = G';\; G^{(2)} = [G', G'];\; \ldots;\; G^{(r+1)} = [G^{(r)}, G^{(r)}].$$

Thus in Examples 17.11 and 17.12 above, $G^{(2)} = \{1\}$.

Proposition 17.14 *Each group $G^{(n)}$ is a normal subgroup of G.*

Proof The proof is by induction on n. When $n = 0$, the result follows trivially. Suppose therefore that $G^{(r)}$ is a normal subgroup of G. It follows by Proposition 17.9 that $G^{(r+1)}$ is a characteristic subgroup of $G^{(r)}$ and so, by Proposition 16.8, is a normal subgroup of G, as required. □

We are now in a position to give some equivalent definitions of solubility.

Proposition 17.15 *The following conditions on a group G are equivalent:*

(i) *G is soluble;*

(ii) *G has a subnormal series*

$$G = G_0 \geq G_1 \geq G_2 \geq \cdots \geq G_r = \{1\}$$

with the quotient groups $G_0/G_1, G_1/G_2, \ldots, G_{r-1}/G_r$ *abelian;*

(iii) *there is an integer* n *for which* $G^{(n)} = \{1\}$.

Proof (i) $\Rightarrow$ (ii). This implication is clear since a normal series is a subnormal series.

(ii) $\Rightarrow$ (iii). We prove, by induction on k, that if G has a subnormal series

$$G = G_0 \geq G_1 \geq G_2 \geq \cdots \geq G_r = \{1\}$$

with abelian quotient groups, then $G^{(k)} \leq G_k$. This will clearly enable us to deduce that $G^{(r)} = \{1\}$. When $k = 0$, $G = G^{(0)} = G_0$. We therefore suppose that $G^{(k)} \leq G_k$. Since G_k/G_{k+1} is abelian, Proposition 17.9 gives that $(G_k)' \leq G_{k+1}$. It follows that

$$G^{(k+1)} = (G^{(k)})' \leq (G_k)' \leq G_{k+1},$$

as required.

(iii) $\Rightarrow$ (i). Finally, if $G^{(n)} = \{1\}$, then the derived series

$$G = G^{(0)} \geq G^{(1)} \geq \cdots \geq G^{(n)} = \{1\}$$

is a normal series for G with abelian quotient groups. □

We now come to the final result of this section, which shows that the class of soluble groups is closed under 'extensions'.

Proposition 17.16 *Let* G *be a group with a normal subgroup* N *such that* N *and* G/N *are soluble groups. Then* G *is a soluble group.*

Proof Since G/N is soluble, there is a subnormal series

$$G/N = G_0/N \geq G_1/N \geq G_2/N \geq \cdots \geq G_r/N = N/N = \{1\}$$

in which the quotients

$$(G_i/N)/(G_{i+1}/N) \cong G_i/G_{i+1} \quad \text{for } 0 \leq i \leq r-1$$

are abelian. Since N is soluble, by Proposition 17.15 there is a subnormal series

$$N = N_0 \geq N_1 \geq N_2 \geq \cdots \geq N_s = \{1\}$$

with abelian quotients. To show that G is soluble, join together these series to obtain a subnormal series

$$G = G_0 \geq G_1 \geq \cdots \geq G_r = N = N_0 \geq N_1 \geq \cdots \geq N_s = \{1\}$$

for G. It is easily checked that the quotient groups in this series are abelian, and so G is soluble by Proposition 17.15. □

Remark The analogue of Proposition 17.16 does not hold for abelian groups. The group $D(3)$ is a non-abelian group with an abelian normal subgroup N with G/N also abelian.

Summary for Chapter 17

A group G is said to be *soluble* if G has a normal series of subgroups with abelian quotient groups. We show that subgroups and quotient groups of soluble groups are soluble. For a finite soluble group, it is shown that any composition factor is cyclic of prime order and that every chief factor is an elementary abelian p-group for some prime p. The *commutator subgroup* of a group G is the subgroup generated by all elements of the form $xyx^{-1}y^{-1}$. This subgroup can be characterised as the smallest normal subgroup of G with abelian quotient group. It is also shown that a group G is soluble if and only if the chain of subgroups obtained by taking successive commutator subgroups terminates in the trivial subgroup.

The use of the word 'soluble' arises from the origins of group theory. Galois considered the problem of when a polynomial equation could be solved using 'radicals'. He was looking for formulae built up from the operations $+, -, \times, \div$ and nth roots which would enable the zeros of a polynomial to be expressed in terms of the coefficients. Solutions of this type were known for quadratic, cubic and quartic equations. The Norwegian mathematician Abel (from whom we obtain the adjective 'abelian') had shown that no such formula exists for the solution of a general equation of degree five.

The technique developed by Galois to solve this problem was to associate a group to each polynomial, and to show that the group is soluble precisely when the polynomial can be solved. It is possible to find a polynomial of degree five whose associated group is $S(5)$. The fact that $S(5)$ is not soluble can then be seen as the reason why there is no formula for the solution of a quintic equation.

Soluble groups have received much attention over the years. The most important fact about finite soluble groups is Hall's generalisation of the Sylow theorems. For any collection π of primes dividing $|G|$, a Hall π-subgroup of G is a subgroup whose order only has primes which belong to the list π and whose index is not divisible by any primes in π. Thus, when π consists of a single prime p, a Hall $\{p\}$-subgroup is simply a Sylow

p-subgroup. Hall proved in 1929 that each finite soluble group has Hall π-subgroups for all sets π of primes. The converse is also true: if G has Hall π-subgroups for all π, then G is soluble. A full account of the theory of Hall subgroups of finite soluble groups may be found in Rose (1978).

Exercises 17

1. Show that the symmetric group $S(n)$ is soluble for $n \leq 4$ but is not soluble for $n \geq 5$.

2. Let $\vartheta : G \to H$ be a group homomorphism. Prove that

(a) $(\vartheta(G))' = \vartheta(G')$;
(b) if H is soluble and ϑ is injective then G is soluble; and
(c) if G is soluble and ϑ is surjective then H is soluble.

3. Let G be the set of maps from $\mathbf{R}$ to $\mathbf{R}$ of the form $\vartheta_{a,b}$ (with $a, b \in \mathbf{R}$ and $a \neq 0$) where $\vartheta_{a,b}(x) = ax + b$. Show that G is a group under composition of maps and that the subset $\{\vartheta_{1,e} : e \in \mathbf{R}\}$ is an abelian subgroup of G. By considering commutators, or otherwise, deduce that G is a soluble group.

4. For any elements x, y of a group G denote by x^y the product yxy^{-1}. Let a, b and c be elements of a group G. Prove that

(a) $[ab, c] = [b, c]^a[a, c]$; and
(b) $[a, bc] = [a, b][a, c]^b$.

5. Given that the symmetric group $S(n)$ has generators $t_1, t_2, \ldots, t_{n-1}$ with relations

$$\begin{aligned} {t_i}^2 &= 1 \quad \text{for } 1 \leq i \leq n-1 \\ (t_i t_{i+1})^3 &= 1 \quad \text{for } 1 \leq i \leq n-2 \\ (t_i t_j)^2 &= 1 \quad \text{for } 1 \leq i, j \leq n-1 \text{ and } |i-j| \geq 2, \end{aligned}$$

find the commutator quotient group of $S(n)$.

6. Let G be the set of all real 4×4 matrices of the form

$$\begin{pmatrix} 1 & a & b & c \\ 0 & 1 & 0 & d \\ 0 & 0 & 1 & e \\ 0 & 0 & 0 & 1 \end{pmatrix}.$$

Find a formula for the product of two elements of G and find the inverse of an element of G. Deduce that G is a group with respect to matrix multiplication. Let A be the subset of matrices in G of the form

$$\begin{pmatrix} 1 & 0 & 0 & c \\ 0 & 1 & 0 & 0 \\ 0 & 0 & 1 & 0 \\ 0 & 0 & 0 & 1 \end{pmatrix}.$$

Show that A is a normal abelian subgroup of G. Prove that G' is contained in A and deduce that G is soluble.

18

Examples of soluble groups

In this chapter, we shall consider further examples of soluble groups. In fact, we shall show that any finite p-group is soluble and also show that any group with fewer than 60 elements is soluble. In proving that a finite p-group is soluble, we shall discuss a stronger property, that of *nilpotence* of finite p-groups.

Definition 18.1 *For any group G, let $Z(G) = Z_1(G)$ denote the centre of G, and $Z_0(G)$ denote the subgroup $\{1\}$. Let $Z_2(G)$ be the subgroup of G defined by $Z_2(G)/Z(G) = Z(G/Z(G))$, and in general let $Z_{i+1}(G)$ be defined by*

$$Z_{i+1}(G)/Z_i(G) = Z(G/Z_i(G)).$$

The series

$$\{1\} \leq Z(G) \leq Z_2(G) \leq \cdots \leq Z_i(G) \leq \cdots$$

is the upper central series *for G.*

Example 18.2 Let G be the dihedral group $D(3)$. We have already seen that $Z(G)$ is the identity subgroup, from which it follows that every term in the upper central series is $\{1\}$.

Proposition 18.3 *A finite p-group is soluble. Every chief factor of a finite p-group is of order p.*

Proof Let G be a finite p-group. We know from Proposition 10.20 that the centre $Z(G)$ is non-trivial. Since each of the quotients $G/Z_i(G)$ is a finite p-group, the upper central series for G

$$\{1\} < Z(G) < Z_2(G) < \cdots < Z_i(G) < \cdots$$

is a strictly increasing series. The fact that G is finite means that this series must terminate, and the only possibility is for G to equal $Z_r(G)$ for some r, giving a series

$$\{1\} < Z(G) < Z_2(G) < \cdots < Z_r(G) = G.$$

This is a normal series for G since the centre of any group is a normal subgroup of the group. The centre of any group is an abelian subgroup, and so the quotient groups $Z_{i+1}(G)/Z_i(G)$ are abelian, showing that G is soluble. Any subgroup of $Z(G)$ is a normal subgroup of G, so refining the series

$$\{1\} < Z(G) < Z_2(G) < \cdots < Z_r(G) = G$$

to a composition series for G will give a series in which each term is in fact a normal subgroup of G. Hence we have a chief series for G which is also a composition series. Since G is soluble, it follows by Proposition 17.6 that the chief factors for G are cyclic of order p. □

Definition 18.4 *A group G is* nilpotent *if the upper central series terminates in G. If the upper central series terminates after r steps, so that $Z_r(G) = G$ but $Z_{r-1}(G) \neq G$, we say that G is* nilpotent of class r.

Remark Proposition 18.3 shows that a finite p-group is nilpotent. Since the upper central series is a normal series, any nilpotent group is soluble.

We now give some properties of finite nilpotent groups which lead to an important characterisation of these groups.

Proposition 18.5 *Let G be a nilpotent group and H be any subgroup of G other than G itself. Then H is a proper subgroup of $N_G(H)$.*

Proof Suppose that G is nilpotent of class r, and let n be the maximal integer such that $Z_n(G) \leq H$. Since H is a proper subgroup of G, $n \leq r-1$. Let g be any element of $Z_{n+1}(G)$ not in H. We show that g normalizes H, so that $H \langle N_G(H)$. To see this, note that, by definition, $Z_{n+1}(G)/Z_n(G)$ is the centre of $G/Z_n(G)$. In particular, the coset $gZ_n(G)$ will commute with each coset $hZ_n(G)$, for all $h \in H$. Thus $ghg^{-1}Z_n(G) = hZ_n(G)$, so that $ghg^{-1} = hz$ for some $z \in Z_n(G)$. Since $Z_n(G) \leq H$, we see that $ghg^{-1} \in H$, so that g normalizes H, as required. □

Definition 18.6 *A subgroup H of a group G is* maximal *if H is a proper subgroup of G and no subgroup of G lies properly between H and G:*

if K is a subgroup of G with $H \leq K \leq G$ then K is either H or G.

Corollary 18.7 *A maximal subgroup of a nilpotent group is normal.*

Proof Let G be a nilpotent group and let H be a maximal subgroup of G. By Proposition 18.5, $H \langle N_G(H)$, and so the maximality of H forces $N_G(H)$ to equal G, so that H is a normal subgroup of G. □

It now is possible to give a natural characterisation of finite nilpotent groups.

Proposition 18.8 *A finite group G is nilpotent if and only if G is an internal direct product of its Sylow subgroups, so that*

$$G = P_1 \times P_2 \times \cdots \times P_r,$$

where $p_1, p_2, \ldots, p_r$ are the distinct primes dividing $|G|$ and P_i is the Sylow p_i-subgroup of G $(1 \leq i \leq r)$.

Proof Suppose that G is the internal direct product of its Sylow subgroups. By definition, each P_i is a normal subgroup of G. For any normal subgroup N of G, it follows by Proposition 11.14 and the Correspondence Theorem that P_iN/N is a normal Sylow p_i-subgroup of G/N. Thus, by Proposition 14.3, G/N is also an internal direct product of its Sylow subgroups. It may easily be checked that

$$Z(P_1 \times P_2 \times \cdots \times P_r) = Z(P_1) \times Z(P_2) \times \cdots \times Z(P_r).$$

Taking N to be $Z(G)$ and using the fact that $Z(G) > \{1\}$, it follows by induction that G is nilpotent.

Conversely, suppose that G is nilpotent, and let p be a prime dividing $|G|$. Let P be a Sylow p-subgroup of G, and let N denote the normaliser of P in G. Since all Sylow p-subgroups of N are conjugate, and P is a normal subgroup of N, it follows that P is the unique Sylow p-subgroup of N. Thus P is the only subgroup of N with $|P|$ elements and so P is a characteristic subgroup of N. Hence P is a normal subgroup of $N_G(N)$ by Proposition 16.8. It follows, therefore, that $N = N_G(N)$, and so $N = G$ by Proposition 18.5. Thus P is a normal subgroup of G. This is true for all primes dividing $|G|$, and so G is an internal direct product of its Sylow subgroups, by Proposition 14.3. □

Corollary 18.9 *Every subgroup and every quotient group of a finite nilpotent group is nilpotent.*

Proof Let G be a finite nilpotent group, so that

$$G = P_1 \times P_2 \times \cdots \times P_r,$$

where P_i is the Sylow p_i-subgroup of G ($1 \leq i \leq r$), and $p_1, p_2, \ldots, p_r$ are the distinct primes dividing $|G|$. For any subgroup H of G, $H \cap P_i$ is a p_i-subgroup of H, and

$$|H : H \cap P_i| = |HP_i : P_i|$$

by Proposition 5.18, so that p_i does not divide $|H : H \cap P_i|$. It follows that $H \cap P_i$ is a Sylow p_i-subgroup of H. Since P_i is a normal subgroup of G, each conjugate of an element of $H \cap P_i$ by an element of H is in $H \cap P_i$. Thus, each Sylow subgroup of H is normal, so that H is an internal direct product of its Sylow subgroups by Proposition 14.3. It then follows by Proposition 18.8 that H is nilpotent.

Similarly, if N is a normal subgroup of G, it can be checked, as in the proof of Proposition 18.8, that P_iN/N is a normal Sylow p_i-subgroup of G/N, so that G/N is nilpotent. □

Remark The conclusions of Corollary 18.9 hold for any nilpotent group, but a different proof is required if the group is not finite.

The following result is another example of the use of the characterisation of Proposition 18.8.

Proposition 18.10 *Let N be a non-trivial normal subgroup of a finite nilpotent group G. Then the intersection $N \cap Z(G)$ has order greater than 1.*

Proof First suppose that G is a p-group. The normal subgroup N is a union of conjugacy classes of G:

$$N = \mathcal{C}_1 \cup \mathcal{C}_2 \cup \cdots \cup \mathcal{C}_r.$$

However, by the Orbit–Stabiliser Theorem, the number of elements in a conjugacy class divides $|G|$, so it is a power of p. Since the conjugacy class containing the identity element consists of one element, there must be other conjugacy classes consisting of one element, so that N contains non-trivial elements of $Z(G)$.

Now suppose that G is an arbitrary nilpotent group, so that G is a direct product of its Sylow subgroups $P_1, P_2, \ldots, P_s$, with $s \geq 1$. Since N is a normal subgroup, $N \cap P_i$ is a Sylow p-subgroup of $N (1 \leq i \leq s)$ by Proposition 11.14. Since N is non-trivial, $N \cap P_i > \{1\}$ for some i, and so by the situation already discussed, $N \cap P_i$, and therefore N, contains a non-identity element of $Z(P_i)$. This element will also be in $Z(G)$ which proves the result. □

We now use the facts we know about nilpotent groups to classify non-abelian groups of order p^3.

Proposition 18.11 *Let p be an odd prime. A non-abelian group of order p^3 is isomorphic to one of*

$$\langle x, y : x^p = 1 = y^p,\ z = [x, y], z^p = 1,\ zx = xz,\ zy = yz\rangle, \text{or}$$

$$\langle x, y : x^{p^2} = 1 = y^p, yxy^{-1} = x^{1+p}\rangle.$$

Proof Let P be a non-abelian group of order p^3. The centre of P is non-trivial, and since P is non-abelian, $Z(P)$ is not P. If $Z(P)$ had order p^2, then $P/Z(P)$ would have order p and so would be cyclic. Then G would be a cyclic extension of a central subgroup and so, by Proposition 10.21, G would be abelian, contrary to hypothesis. It follows that $Z(P)$ is of order p, say $P = \langle z\rangle$. The quotient group $P/Z(P)$ has order p^2, but is not cyclic since P is non-abelian and so cannot be a cyclic extension of a central subgroup. Hence $P/Z(P)$ is isomorphic to $C_p \times C_p$.

Case 1. Suppose that every element of G has order p. Let x and y be elements of P such that the cosets $xZ(P)$ and $yZ(P)$ generate $P/Z(P)$. Since every element of P has order p, $x^p = 1 = y^p$. The group $P/Z(P)$ is abelian, so $[x, y] \in Z(P) = \langle z\rangle$. If $[x, y] = 1$, then x and y would commute and since z is central, any two elements of the form $x^i y^j z^k$ would commute. But all such elements are distinct: if $x^i y^j z^k = x^a y^b z^c$, then the corresponding elements in $P/Z(P)$ are equal, so $i = a$ and $j = b$, and so it follows that $k = c$. This shows that P would be abelian, and so we deduce that the commutator $[x, y]$ is not 1. Thus $[x, y] = z^r$ for some non-zero r. An easy proof by induction on i shows that $[x^i, y] = z^{ri}$. So if we replace x by x^s, where s is the inverse of r modulo p, P has presentation

$$\langle x, y : x^p = 1 = y^p, z = [x, y], z^p = 1, zx = xz, zy = yz\rangle,$$

as required.

Case 2. We may therefore suppose that our non-abelian group of order p^3 has an element x of order p^2. Since $\langle x\rangle$ has order p^2, it is a maximal subgroup of P, and so by Corollary 18.7, $\langle x\rangle$ is a normal subgroup of P. If $Z(P)$ were not contained in $\langle x\rangle$, then P would be $\langle x\rangle Z(P)$. It would then follow that P is an internal direct product of $\langle x\rangle$ and $Z(P)$, and so P would be abelian. Hence $Z(P) < \langle x\rangle$ and so by Proposition 4.14, $Z(P)$ is the unique subgroup of order p in $\langle x\rangle$. We may therefore suppose that $Z(P)$ is generated by the element $z = x^p$. Let y be any element not in $\langle x\rangle$, so that the cosets $xZ(P)$ and $yZ(P)$ generate $P/Z(P)$ and y^p is in $Z(P)$. Since $P/Z(P)$ is abelian, $yxy^{-1}x^{-1} \in Z(P)$, and so $yxy^{-1} = x^{1+kp}$ for some k. Replacing y with y^i where $ik \equiv 1 \bmod p$, we may take k to be 1. Finally, we show that y may be taken to have order p, so that this leads to the presentation

$$\langle x, y : x^{p^2} = 1 = y^p,\ yxy^{-1} = x^{1+p} \rangle.$$

This is done by replacing y by $x^i y$, where $y^p = z^{-i}$. To see this, note that

$$\begin{aligned} (x^i y)^p &= x^i y x^y \ldots x^i y \\ &= x^i (x^i)^y (x^i)^{y^2} \ldots (x^i)^{y^{p-1}} y^p \\ &= x^i (x^i)^y (x^i)^{y^2} \ldots (x^i)^{y^{p-1}} z^{-i} \end{aligned}$$

This is $x^k z^{-i}$, where

$$\begin{aligned} k &= i(1 + (1+p) + (1+p)^2 + \cdots + (1+p)^{p-1}) \\ &= i\frac{(1+p)^p - 1}{(p+1) - 1}. \end{aligned}$$

Since

$$(1+p)^p \equiv 1 + p^2 \mod p^3,$$

we see that $(x^i y)^p = x^{ip} z^{-i} = 1$, as required. □

To conclude this chapter, we provide a further list of examples of soluble groups by showing the following:

Theorem 18.12 *Every group with less than 60 elements is soluble.*

Proof The proof of this fact is a case-by-case analysis, but some general remarks may be made:

(1) all p-groups are nilpotent and therefore soluble;
(2) if p and q are primes with $p > q$, a group G of order pq has a normal Sylow p-subgroup P by Proposition 12.1, and so is soluble since P and G/P are both abelian.
(3) a group of order 12 is soluble as a consequence of Proposition 12.3;
(4) for any primes p, q, groups of order $p^2 q$ are soluble as a consequence of Proposition 12.4;
(5) groups of order 30 are soluble using Proposition 12.5;
(6) groups of order 24 are soluble as a simple consequence of Proposition 12.7.

The integers between 1 and 59 which are not covered by the above list are: 36; 40; 42; 48; 54 and 56. We now consider these in turn.

(7) A group with 40, 42 or 54 elements has a normal Sylow p-subgroup for some divisor p of its order. The number of Sylow 5-subgroups in a group with 40 elements must be 1; the number of Sylow 7-subgroups in a group with 42 elements is also 1. Any Sylow 3-subgroup in a group with 54 elements has index 2 and so is normal. Thus in each of these cases, our group has a soluble normal subgroup with soluble quotient group and so is soluble by Proposition 17.16.

(8) Suppose G is a group with 56 elements. Then if the Sylow 7-subgroup is normal, G is soluble. Otherwise G has 8 Sylow 7-subgroups, whose pairwise intersection is $\{1\}$. It follows that G has 48 elements of order 7 and so can only have one Sylow 2-subgroup Q, say. Thus Q is normal with quotient of order 7, so G is soluble.

(9) Groups of order 48 are analysed in a similar way to those of order 24 (see Proposition 12.7).

(10) Finally, we are left with groups with 36 elements. Again, this is a similar argument to that for groups of order 24, but we present most of the details for the sake of completeness.

Let G have 36 elements. If G has a unique Sylow 3-subgroup, P say, then P is abelian because it has order 9, and G/P has order 4 and so is also abelian. Thus G is soluble in this case. We may therefore suppose that G has 4 Sylow 3-subgroups, and we choose P_1 and P_2 to be distinct of order 9. Thus $|P_1P_2|$ has at most 36 elements, and so $P_1 \cap P_2 > \{1\}$. It follows since P_1 and P_2 are distinct that $P_1 \cap P_2$ has order 3, so that $|P_1P_2| = 9^2/3 = 27$. The subgroup generated by P_1 and P_2 therefore contains at least 27 elements, and so this subgroup must equal G. Since P_1 and P_2 are abelian, $P_1 \cap P_2$ is a normal subgroup of both P_1 and P_2, and therefore of the subgroup generated by them. Thus G has a normal subgroup of order 3 in this case with the quotient group being of order 12. The result follows since groups of order 12 are soluble. □

Summary for Chapter 18

In this chapter, we show that any finite p-group is soluble. In fact any finite p-group is shown to be a *nilpotent* group. The finite nilpotent groups are those which are a direct product of their Sylow subgroups. Among the properties of finite nilpotent groups which are established are the facts that a maximal subgroup of a nilpotent group is normal, and also that any normal subgroup of a nilpotent group contains non-identity central elements.

It is also shown that any group with fewer than 60 elements is soluble, so that 60 is the order of the smallest possible non-abelian simple group. We have already seen that there is a non-abelian simple group with 60 elements.

Exercises 18

1. Show that the class of nilpotent groups is not closed under extensions: there is a non-nilpotent group G with a normal subgroup N such that N and G/N are both nilpotent.

2. Let G be the group of order 27 consisting of the 3×3 matrices of the form

$$A = \begin{pmatrix} 1 & a & b \\ 0 & 1 & c \\ 0 & 0 & 1 \end{pmatrix},$$

where a, b and c are integers modulo 3, so that each of a, b and c may be taken to be one of 0, 1 or 2 with arithmetical operations modulo 3 (so that, for example, $1 + 2 = 0$ and $2 \times 2 = 1$). Prove that every element of G has order 3. Find the centre of G. Hence find a chief series for G.

3. Let P be the group

$$\langle x, y : x^9 = y^3 = 1, yxy^{-1} = x^4 \rangle,$$

so that P consists of the 27 elements of the form $x^j y^k$ with j being one of $0, 1, \ldots, 8$ and k being 0, 1 or 2. Show that $\langle x \rangle$ is a normal subgroup of P and that x^3 is in the centre of P. Find a chief series for P.

4. Explain why no two of the following 5 groups with 27 elements are isomorphic: C_{27}; $C_9 \times C_3$; $C_3 \times C_3 \times C_3$; the group G in question 2; and the group P in question 3. Find an example of two non-isomorphic groups which have the same number of elements of each possible order.

5. Show that groups of all possible orders between 61 and 71 (inclusive) are soluble.

19

Semidirect products and wreath products

In this chapter, we start to consider the problem of constructing groups. The techniques which we shall develop during the next three chapters will be used in Chapter 22 to construct groups of certain orders. One objective of these results is to determine the complete list of isomorphism classes of groups of order less than 32. We first discuss the idea of a semidirect product and consider some of its applications.

Definition 19.1 *A group G is* a semidirect product *of a subgroup N by a subgroup H if the following conditions are satisfied:*

(i) $G = NH$;
(ii) *N is a normal subgroup of G; and*
(iii) $H \cap N = \{1\}$.

Remark If G is the semidirect product of N by H, the First Isomorphism Theorem, Theorem 8.15, shows that

$$G/N = NH/N \cong H/(H \cap N) = H/\{1\} \cong H.$$

We say that G is an extension of N by H in this case. In general, when we say that G is an extension of A by B, B need not be a subgroup of G, but B is isomorphic to the quotient group G/A. We shall discuss extensions in general in the next chapter.

Example 19.2 Any (internal) direct product of groups H, N is a semidirect product using Proposition 13.5, since $H \cap N = \{1\}$ and H, N are both normal subgroups of G.

Example 19.3 The dihedral group

$$D(4) = \langle x, y : x^4 = 1 = y^2,\ y^{-1}xy = x^{-1} \rangle$$

is a semidirect product of $\langle x \rangle$ by $\langle y \rangle$. This group is not, however, a direct product of these two subgroups since $\langle y \rangle$ is not a normal subgroup of G.

The quaternion group

$$Q_8 = \langle a, b : a^4 = 1,\ a^2 = b^2,\ b^{-1}ab = a^{-1} \rangle$$

is not a semidirect product of a group of order 4 by a group of order 2. This is because Q_8 has a unique element of order 2 and so only has one subgroup $\langle a^2 \rangle$ of order 2. It follows that each subgroup of Q_8 with 4 elements must be cyclic and so have a^2 as its element of order 2. Since each subgroup with 4 elements contains the only subgroup with 2 elements, condition (iii) of the definition of semidirect product fails.

Proposition 19.4 *Let G be a semidirect product of N by H. For each element h of H, the map $\vartheta_h : N \to N$ defined by $\vartheta_h(n) = hnh^{-1}$ is an automorphism of N. The map $\vartheta : H \to \mathrm{Aut}(N)$ defined by $\vartheta(h) = \vartheta_h$ is a homomorphism.*

Proof Since N is a normal subgroup of G, hnh^{-1} is an element of N. Then ϑ_h is a homomorphism since

$$\vartheta_h(n_1 n_2) = hn_1 n_2 h^{-1} = hn_1 h^{-1} hn_2 h^{-1} = \vartheta_h(n_1)\vartheta_h(n_2).$$

Also ϑ_h is injective since if $\vartheta_h(n_1) = \vartheta_h(n_2)$ then $hn_1h^{-1} = hn_2h^{-1}$, from which it follows that $n_1 = n_2$. Finally, ϑ_h is surjective since, for any $n \in N$, $\vartheta_h(h^{-1}nh) = n$.

To show that ϑ is a homomorphism, consider

$$\begin{aligned} \vartheta(h_1h_2)(n) &= (h_1h_2)n(h_1h_2)^{-1} \\ &= h_1(h_2nh_2^{-1})h_1^{-1} \\ &= \vartheta(h_1)(h_2nh_2^{-1}) \\ &= \vartheta(h_1)\vartheta(h_2)(n), \end{aligned}$$

as required. □

We now consider the converse problem of how to construct a semidirect product.

Proposition 19.5 *Given any groups N and H, and a homomorphism from H to* Aut(N) *for which the image of h is denoted ϑ_h, let G be the set of ordered pairs $\{(n,h) : n \in N, h \in H\}$. Then G is a group under the multiplication defined by*

$$(n_1,h_1)(n_2,h_2) = (n_1\vartheta_{h_1}(n_2),\ h_1h_2).$$

The sets $N_0 = \{(n,1) : n \in N\}$ and $H_0 = \{(1,h) : h \in H\}$ are subgroups of G isomorphic to N and H, respectively, and G is a semidirect product of N_0 by H_0.

Proof We first check that G is a group, the closure axiom being obvious. For associativity,

$$\begin{aligned}((n_1,h_1)(n_2,h_2))(n_3,h_3) &= ((n_1\vartheta_{h_1}(n_2),\ h_1h_2))(n_3,h_3)\\ &= (n_1\vartheta_{h_1}(n_2)\vartheta_{h_1h_2}(n_3),\ (h_1h_2)h_3)\\ &= (n_1\vartheta_{h_1}(n_2\vartheta_{h_2}(n_3)), h_1(h_2h_3))\\ &= (n_1,h_1)(n_2\vartheta_{h_2}(n_3), h_2h_3)\\ &= (n_1,h_1)((n_2,h_2)(n_3,h_3)),\end{aligned}$$

as required. The identity element is $(1,1)$ since

$$(n,h)(1,1) = (n\vartheta_h(1),h) = (n,h) = (1\vartheta_1(n),h) = (1,1)(n,h).$$

Since

$$(n,h)(\vartheta_h^{-1}(n^{-1}),h^{-1}) = (1,1),$$

each element of G has a right inverse. This is in fact sufficient to show that each element of G has a two-sided inverse. To see this, note that for each g in G, there is an element g' such that $gg' = 1$. Thus there is an element h such that $g'h = 1$, then

$$1 = g'h = g'1h = g'gg'h = g'g1 = g'g,$$

so that g' is also a left inverse of g. We have therefore proved that

$$(\vartheta_h^{-1}(n^{-1},h^{-1}))$$

is the (two-sided) inverse of (n,h), and that G is a group.

Since $(1,h_1)(1,h_2) = (1,h_1h_2)$, H_0 is a subgroup of G isomorphic to H. Also $(n_1,1)(n_2,1) = (n_1n_2,1)$, so N_0 is a subgroup isomorphic to N.

Since $(n, h) = (n, 1)(1, h)$, we see that $G = N_0 H_0$. Finally, N_0 is a normal subgroup of G since

$$(n, h)(n_1, 1)(n, h)^{-1} = (n_0, 1)$$

for some $n_0 \in N$. This completes the proof that G is a semidirect product of N_0 by H_0. □

Example 19.6 One natural example of the construction of Proposition 19.5 arises when H is equal to $\mathrm{Aut}(N)$ and the map ϑ is the identity map from $H(= \mathrm{Aut}(N))$ to $\mathrm{Aut}(N)$ so that $\vartheta(\phi) = \phi$. The resulting semidirect product is known as the *holomorph* of N. As an example, if N is cyclic of order 3, say $N = \langle x : x^3 = 1 \rangle$, an automorphism of N is determined by its value on x. Since an automorphism is injective, it cannot map x to 1 and so there are two automorphisms of N (specified by $x \to x$ and $x \to x^2$). The holomorph of N therefore has six elements and is non-abelian, so it is isomorphic to the dihedral group $D(3)$.

Example 19.7 Consider all possible semidirect products G of $N = C_3$ by $H = C_2$. Any such group will have six elements. We already know that any group with six elements is isomorphic either to C_6 or to $D(3)$, both of these groups being semidirect products. We now use our theory of semidirect products to re-establish this fact. According to Proposition 19.5, we need to associate an automorphism of C_3 to the elements 1 and x of C_2. Since the automorphism associated to 1 must be the identity, we only have one choice to make. Any automorphism of C_3 is determined by its effect on the generator, y. Since y cannot map to 1, the only non-trivial automorphism maps y to y^{-1} (and hence maps y^2 to $y^4 = y$). When ϑ_x is also the identity map the group constructed in Proposition 19.4 has multiplication

$$(n_1, h_1)(n_2, h_2) = (n_1 n_2, h_1 h_2),$$

so that G is the direct product of C_2 and C_3, which is isomorphic to C_6 by Proposition 13.1.

When ϑ_x inverts y, the multiplication becomes

$$(n_1, h_1)(n_2, h_2) = (n_1 \vartheta_{h_1}(n_2), h_1 h_2) = (n_1 n_2^{-1}, h_1 h_2).$$

Thus the six elements of G are all of the form

$$(y^i, x^j) = (y^i, 1)(1, x^j)$$

Also $(y, 1)$ has order 3, $(1, x)$ has order 2, and

$$(1,x)(y,1) = (\vartheta_x(y),x) = (y^2,x) = (y,1)^2(1,x),$$

so that G is isomorphic to the dihedral group $D(3)$ in this case.

Remark As with internal direct products, it is often convenient to drop the ordered pair notation, and write elements of a semidirect product of N by H by juxtaposing elements of N and H. In this notation the definition of multiplication becomes

$$n_1h_1n_2h_2 = n_1\vartheta_{h_1}(n_2)h_1h_2.$$

We next turn our attention to wreath products. These are examples of semidirect products. However, the situation is complicated here by the fact that the phrase 'wreath product' is used in the literature for two *different* constructions. We shall attempt to clarify the position by introducing these separately, and giving each its own notation.

Definition 19.8 *Let G and H be any two finite groups. The* regular wreath product G rwr H *is a semidirect product of the group N by H, where N is $G^{|H|}$, the direct product of $|H|$ copies of G. Thus, if $|H| = n$, the elements of N are n-tuples of the form $(g_1, g_2, \ldots, g_n)$ with each g_i in G. Now to specify the extension, choose some fixed ordering $h_1, h_2, \ldots, h_n$ of the elements of H. The automorphism ϑ_h of G^n associated with an element h of H is then defined by*

$$\vartheta_h(g_1, g_2, \ldots, g_n) = (g_{\pi(1)}, g_{\pi(2)}, \ldots, g_{\pi(n)}),$$

where π is the permutation of $\{1, \ldots, n\}$ defined by $hh_i = h_{\pi(i)}$. It may be checked that $\vartheta_h\vartheta_k = \vartheta_{hk}$. Thus the map $H \to \mathrm{Aut}(G^n)$ determined by $h \mapsto \vartheta_h$ is a homomorphism. The semidirect product is therefore well-defined. Note that the number of elements in G rwr H is $n|G|^n$, where $n = |H|$.

Example 19.9 The wreath product C_2 rwr C_2 has order 8, and this group is a semidirect product of $C_2 \times C_2$ by C_2. In fact, this group is isomorphic to the dihedral group $D(4)$. To see this, we apply the above definition of the wreath product to a group G of order 2 generated by x, and let N be the group $G^{(2)} = G \times G$ with 4 elements $(1,1), (1,x), (x,1)$ and (x,x). Let h be a generator for a cyclic group H of order 2. Let the fixed ordering of H be $\{h_1 = 1, h_2 = h\}$. Then ϑ_1 is the identity automorphism and, by definition, ϑ_h is given by

$$\vartheta_h(g_1, g_2) = (g_2, g_1)$$

so that ϑ_h fixes both $(1,1)$ and (x,x) and ϑ_h interchanges the other two elements of N. Thus, using Proposition 19.5 and writing the semidirect product using juxtaposition,

$$((1,x)h)^2 = (1,x)\vartheta_h(1,x)h^2 = (1,x)(x,1) = (x,x),$$

so that $(1,x)h$ has order 4. Also

$$((1,x)h)^{-1} = h^{-1}(1,x)^{-1} = h(1,x) = h((1,x)h)h^{-1},$$

so writing $(1,x)h$ as a and h as b, C_2 rwr C_2 has presentation

$$\langle a,b : a^4 = 1 = b^2, bab = a^{-1}\rangle,$$

and so is isomorphic to $D(4)$, as required.

One of the standard uses of this regular wreath product construction lies in the determination of the structure of the Sylow p-subgroups of the symmetric groups.

Proposition 19.10 *Let p be a prime integer, and let k be any positive integer. The Sylow p-subgroup of the symmetric group $S(p^k)$ is the iterated wreath product (with k copies of C_p)*

$$(\ldots((C_p \text{ rwr } C_p) \text{ rwr } C_p)\ldots \text{rwr } C_p).$$

Proof The proof is by induction on k. Note that when $k = 1$, the highest power of p dividing $p!$ is p, so that the Sylow p-subgroup is indeed cyclic of order p. We therefore suppose that the result holds for the symmetric group of degree p^k. The first objective is to determine the highest power of p dividing $(p^n)!$. The number of the integers $1, \ldots, p^n$ divisible by p is clearly p^{n-1}. Of these p^{n-1} integers, precisely p^{n-2} are divisible by p^2, and so on. Thus the highest power of p dividing $(p^n)!$ is $p^{r(n)}$ where

$$r(n) = p^{n-1} + p^{n-2} + \cdots + p + 1.$$

The order of a regular wreath product of k copies of the cyclic group C_p is equal to $p^{r(k)}$. It follows that the regular wreath product of $k+1$ copies of C_p has the same order (namely $(p^{r(k)})^p \times p$) as the order of the Sylow p-subgroup of $S(p^{k+1})$, since

$$pr(k) + 1 = p(p^{k-1} + p^{k-2} + \cdots + p + 1) + 1 = r(k+1).$$

It only remains to show that this wreath product occurs as a subgroup of $S(p^{k+1})$. In $S(p^{k+1})$, let N denote the direct product of p copies of

$S(p^k)$, where we regard the first copy of $S(p^k)$ as the symmetric group on $\{1, \ldots, p^k\}$, the second copy as the symmetric group on the numbers $\{1+p^k, \ldots, 2p^k\}$, and so on. These copies of $S(p^k)$ commute because they act on disjoint sets. Let h be the element of $S(p^{k+1})$ consisting of p^k cycles each of length p, these cycles all being of the form

$$(i, i+p^k, i+2p^k, \ldots, i+(p-1)p^k),$$

for $1 \leq i \leq p^k$. By inductive hypothesis, the Sylow p-subgroup of $S(p^k)$ is an iterated wreath product of k copies of C_p. It is clear that the semidirect product of the Sylow p-subgroup of N by $H = \langle h \rangle$ (where h is the element of $S(p^{k+1})$ defined above), is isomorphic to the regular wreath product of $k+1$ copies of C_p, thus completing the proof. □

Corollary 19.11 *Let p be a prime and n be any positive integer. Let*

$$n = a_0 + a_1 p + a_2 p^2 + \cdots + a_k p^k \text{ (with } 0 \leq a_i \leq p-1) \qquad (*)$$

be the expansion of n to the base p. Then each Sylow p-subgroup of the symmetric group $S(n)$ is a direct product

$$(S_1)^{a_1} \times (S_2)^{a_2} \times \cdots \times (S_k)^{a_k},$$

where S_i is a Sylow p-subgroup of the symmetric group $S(p^i)$, so that S_i is the regular wreath product of i copies of the cyclic group C_p.

Proof We first calculate the power of p dividing $n!$. The number of integers between 1 and n divisible by p is $[n/p]$, where [ ] denotes the integer part function. Of these integers a further $[n/p^2]$ are divisible by p^2, and so on. It follows that the power of p dividing $n!$ is p^t, where

$$t = [n/p] + [n/p^2] + [n/p^3] + \cdots$$

Using the expansion $(*)$ of n to the base p, this power can also be written as

$$\begin{aligned} t &= (a_1 + a_2 p + \cdots + a_k p^{k-1}) + (a_2 + \cdots + a_k p^{k-2}) + \cdots \\ &= a_1 + a_2(p+1) + a_3(p^2+p+1) + \cdots, \end{aligned}$$

so that p^t is the order of the group

$$(S_1)^{a_1} \times (S_2)^{a_2} \times \cdots \times (S_k)^{a_k}$$

by Proposition 19.10. It only remains to show that $S(n)$ has a subgroup isomorphic to the given direct product. This is done by dividing the set with n elements into disjoint subsets with a_0 of these of size $1, a_1$ of size $p, \ldots, a_k$ of size p^k, and then applying Proposition 19.10 to each of these disjoint subsets. □

We now turn to the other type of wreath product.

Definition 19.12 *Let G and H be finite groups with H a subgroup of the symmetric group $S(n)$.* The permutation wreath product, *G pwr H, is the semidirect product of a normal subgroup N by H, where N is the direct product of n copies of G. Thus, the elements of N are n-tuples $(g_1, g_2, \ldots, g_n)$ with each g_i in G. The automorphism ϑ_h of G^n associated with a permutation h in H is then defined by*

$$\vartheta_h(g_1, g_2, \ldots, g_n) = (g_{h(1)}, g_{h(2)}, \ldots, g_{h(n)}).$$

Remark 1 In the above construction, H acts by conjugation on N, permuting the n direct factors. Since the map $H \to \mathrm{Aut}(G^n)$ determined by $h \mapsto \vartheta_h$ is easily checked to be a homomorphism, the semidirect product is well-defined. Note that the number of elements in G pwr H is $|H||G|^n$.

Remark 2 The regular wreath product may be regarded as a special case of the permutation wreath product. In this case, H is regarded in its *regular* permutation representation as a subset of $S(|H|)$ in which the permutation π_h associated with an element h on $H = \{h_1, \ldots, h_n\}$ is defined by $\pi_h(h_i) = hh_i$. In general, these two constructions have different orders. For example if $H = S(3)$, then the group G rwr H would have order $6|G|^6$, whereas G pwr H would have order $6|G|^3$.

Example 19.13 An important source of examples of the permutation wreath product arises when the permutation group is taken to be the symmetric group $S(n)$. By way of illustration, we consider C_2 pwr $S(n)$. This is the semidirect product of n copies of the cyclic group C_2 by the group $S(n)$ of order $n!$. In this case the permutation wreath product has order $2^n n!$.

This permutation wreath product C_2 pwr $S(n)$ is also known as the *hyperoctahedral group*. It is a special case of a *generalised symmetric group*, also known as a *monomial group*, and is often considered in a matrix form as follows. The group, $P(n)$, of $n \times n$ permutation matrices, is isomorphic to the symmetric group $S(n)$. To see this, represent $\pi \in S(n)$ by the $n \times n$ matrix with precisely one non-zero entry in the ith column $(1 \leq i \leq n)$, this being an entry with value 1 in the $\pi(i)$th row. Thus when $n = 3$, the permutation matrices are

$$\begin{pmatrix} 1 & 0 & 0 \\ 0 & 1 & 0 \\ 0 & 0 & 1 \end{pmatrix}, \begin{pmatrix} 0 & 0 & 1 \\ 1 & 0 & 0 \\ 0 & 1 & 0 \end{pmatrix}, \begin{pmatrix} 0 & 1 & 0 \\ 0 & 0 & 1 \\ 1 & 0 & 0 \end{pmatrix},$$

$$\begin{pmatrix} 0 & 1 & 0 \\ 1 & 0 & 0 \\ 0 & 0 & 1 \end{pmatrix}, \begin{pmatrix} 0 & 0 & 1 \\ 0 & 1 & 0 \\ 1 & 0 & 0 \end{pmatrix}, \begin{pmatrix} 1 & 0 & 0 \\ 0 & 0 & 1 \\ 0 & 1 & 0 \end{pmatrix},$$

representing the permutations $1, (1\ 2\ 3)$, $(1\ 3\ 2)$, $(1\ 2)$, $(1\ 3)$ and $(2\ 3)$, respectively. These six matrices form a group isomorphic to the symmetric group $S(3)$. In general the $n \times n$ permutation matrices form a group $P(n)$ isomorphic to $S(n)$.

Now let B_n be the group of $n \times n$ matrices in which each non-zero entry of each matrix in $P(n)$ is allowed to be ± 1, so that B_n has a normal subgroup, D, of order 2^n consisting of diagonal matrices with each diagonal entry being either 1 or -1. Since each monomial matrix can be written as a product of a permutation matrix and a diagonal matrix, it may be seen that B_n is the semidirect product of D by the group $P(n)$ of permutation matrices. It follows that this group B_n is isomorphic to the permutation wreath product C_2 pwr $S(n)$.

Similarly the generalised symmetric group is the permutation wreath product C_m pwr $S(n)$. This group also has a matrix interpretation, in which each non-zero entry in the group $P(n)$ is allowed to be a power of the complex number $e^{2\pi i/m}$.

The notation B_n comes from the theory of simple Lie algebras and their Weyl groups. The group B_n has a special type of presentation, which can be described in terms of our matrix representation of the elements of C_2 pwr $S(n)$. For $1 \leq i \leq n-1$, let b_i be the matrix which differs from the identity $n \times n$ matrix only in that there is a 2×2 block

$$\begin{pmatrix} 0 & 1 \\ 1 & 0 \end{pmatrix}$$

in the location determined by rows i and $i+1$ and columns i and $i+1$. Since each b_i transposes adjacent basis vectors, $\langle b_1, b_2, \ldots, b_{n-1} \rangle$ is isomorphic to the symmetric group $S(n)$. Finally, let b_n be the matrix which differs from the identity matrix I_n only in that its entry in the (n, n) location is -1. Thus b_n commutes with b_i for $1 \leq i \leq n-2$, whereas $b_n b_{n-1}$ has the 2×2 matrix

$$\begin{pmatrix} 0 & 1 \\ -1 & 0 \end{pmatrix}$$

in its bottom right-hand corner. Since the square of this 2×2 block is the negative of the identity matrix, $b_n b_{n-1}$ has order 4. The conjugates of b_n under $\langle b_1, b_2, \ldots, b_{n-1} \rangle$ generate the diagonal subgroup D. Thus B_n is generated by $b_1, \ldots, b_{n-1}$ (which generate $P(n)$) and b_n, and it may be shown that it has a presentation:

$$
\begin{aligned}
b_1^2 &= b_2^2 = \cdots = b_n^2 = 1; \\
(b_i b_{i+1})^3 &= 1 \quad \text{for } 1 \le i \le n-2; \\
(b_{n-1} b_n)^4 &= 1; \\
(b_i b_j)^2 &= 1 \quad \text{for } 1 \le i, j \le n-1, |i-j| > 1; \text{ and} \\
(b_i b_n)^2 &= 1 \quad \text{for } 1 \le i \le n-2.
\end{aligned}
$$

For small values of n, we can give alternative descriptions of B_n. Thus, when $n = 2$, the group B_2 has order $2^2 2! = 8$, and B_2 is isomorphic to the dihedral group of order 8. (In fact when H is cyclic and of order n, G rwr H is isomorphic to G pwr H.) When $n = 3$, B_3 has order $2^3 3! = 48$. The centre of this group consists of the matrices I_3 and $-I_3$. There are six matrices in $P(3)$, and eighteen matrices in G with precisely two entries equal to -1. It may be checked that these 24 matrices form a subgroup N. Since N has index 2, it is a normal subgroup of G. Thus B_3 is a direct product $C_2 \times N$. Furthermore, N is itself isomorphic to the symmetric group $S(4)$: define matrices a_1, a_2, and a_3 by

$$
a_1 = \begin{pmatrix} 0 & 1 & 0 \\ 1 & 0 & 0 \\ 0 & 0 & 1 \end{pmatrix}; \; a_2 = \begin{pmatrix} 1 & 0 & 0 \\ 0 & 0 & 1 \\ 0 & 1 & 0 \end{pmatrix}; \; a_3 = \begin{pmatrix} 0 & -1 & 0 \\ -1 & 0 & 0 \\ 0 & 0 & 1 \end{pmatrix}.
$$

It may be checked that these matrices are in the group and also satisfy the relations:

$$
a_1^2 = a_2^2 = a_3^2 = 1, \; (a_1 a_2)^3 = 1, \; (a_1 a_3)^2 = 1, \text{ and } (a_2 a_3)^3 = 1.
$$

Since $\langle a_1, a_2, a_3 \rangle$ has 24 elements, this group is isomorphic to $S(4)$, as required.

Remark Yet another situation in which these groups of type B_n occur is as the centraliser of an element of cycle type 2^n in the symmetric group $S(2n)$. So the centraliser of the permutation $(1\ 2)(3\ 4)\ldots(2n-1\ 2n)$ in $S(2n)$ is C_2 pwr $S(n)$.

Summary for Chapter 19

We say that G is a *semidirect product* of N by H if

$G = NH$;

N is a normal subgroup of G; and

$N \cap H = \{1\}$.

We also show that a semidirect product may be constructed from groups H and N by taking a homomorphism ϑ from H to $\mathrm{Aut}(N)$ and defining multiplication for G by

$$n_1 h_1 n_2 h_2 = n_1 \vartheta_{h_1}(n_2) h_1 h_2.$$

An important illustration of this situation occurs when H is equal to $\mathrm{Aut}(N)$, and ϑ is the trivial automorphism, giving the *holomorph* of N.

Other important examples of semidirect products occur in the construction of *wreath products*. There are two distinct constructions covered by this term. In the first of these, which we refer to as the regular wreath product, the group G rwr H is a semidirect product of $|H|$ copies of G by H. One of the earliest uses of this wreath product, given by Kaloujnine (1948), was to describe the structure of the Sylow p-subgroups of the symmetric group $S(n)$ (see Corollary 19.11). In the other type of wreath product, the permutation wreath product, the group H needs to be a subgroup of some symmetric group, $S(n)$, say. Then G pwr H is a semidirect product of n copies of G by H. An important example of this occurs in the groups *of type* B_n, which are the groups C_2 pwr $S(n)$.

Exercises 19

1. Let each of G, H and K be finite non-trivial groups. Explain why the wreath product (G rwr (H rwr K)) cannot possibly be isomorphic to the wreath product ((G rwr H) rwr K).

2. Show that the centraliser of $(1\ 2)(3\ 4)(5\ 6)$ in $S(6)$ is isomorphic to the wreath product C_2 pwr $S(3)$.

3. Let G be a group with precisely two distinct prime divisors, p and q. Given that the Sylow p-subgroup is a normal subgroup of G, show that G is a semidirect product.

4. Let G be a semidirect product of a group N with two elements by a subgroup H. Show that G is an internal direct product of N and H.

5. Construct all semidirect products of C_3 by C_3.

20

Extensions

In this chapter, we shall generalise the idea of a semidirect product and consider general extensions of one group by another.

Definition 20.1 *A group G is an* extension of N by H *if G has a normal subgroup N such that the quotient group G/N is isomorphic to H.*

Remark Suppose that G is a semidirect product of N by H. The First Isomorphism Theorem then shows that

$$G/N = HN/N \cong H/H \cap N \cong H,$$

so that G is an extension of N by H.

Example 20.2 The symmetric group $S(5)$ is an extension of $A(5)$ by C_2 since $A(5)$ is a normal subgroup of index 2. Taking H to be the subgroup $\{1, (1\ 2)\}$, we see that $S(5)$ is in fact a semidirect product of $A(5)$ by C_2.

Example 20.3 Let G be a cyclic group of order 4 generated by x. The subgroup generated by x^2 is a normal cyclic subgroup of order 2, and the quotient group is cyclic of order 2. By definition, therefore, G is an extension of C_2 by C_2. However, the subgroup x^2 is the unique subgroup of order 2, so this extension is not a semidirect product.

In the first part of this chapter, we shall investigate how a given extension G of N by H is built up. This analysis will motivate later discussions, when we consider the reverse situation where we are given N and H and wish to find all (isomorphism classes of) groups which are extensions of N by H.

Definition 20.4 *Let G be an extension of N by H with $\phi : H \to G/N$ an isomorphism. A* section *of G through H is any set $\{s(h) : h \in H\}$ of elements of G such that:*

(i) $s(1) = 1$; *and*

(ii) *$s(h)$ is a representative for the right coset $\phi(h)$, so that $\phi(h) = Ns(h)$.*

Remark If G is a semidirect product of N by H, the elements of H are a section of G through H which is a subgroup.

Example 20.5 For the extension $S(5)$ of $A(5)$ by C_2, any pair of elements of $S(5)$ of the form $\{1, \pi\}$, where π is any odd permutation of $S(n)$, is a section of $S(5)$ through C_2. In particular, $\{1, (1\ 2)\}$ is a section which happens also to be a subgroup of $S(5)$.

Remark Let G be an extension of N by H with $\phi : H \to G/N$ an isomorphism. Since ϕ is an isomorphism, for any elements h_1 and h_2 of H, $s(h_1)s(h_2)$ is in the same right coset as $s(h_1h_2)$, and so there exists an element $f(h_1, h_2)$ in N such that

$$s(h_1)s(h_2) = f(h_1, h_2)s(h_1h_2).$$

Definition 20.6 *Let G be an extension of N by H with $\{s(h) : h \in H\}$ a section of G through H. The map $f : H \times H \to N$ defined by*

$$f(h_1, h_2) = s(h_1)s(h_2)(s(h_1h_2))^{-1},$$

for all h_1, h_2 in H, is the sectional factor set for the extension G *with section $\{s(h) : h \in H\}$.*

Remark It is clear that in general the sectional factor set depends on the choice of section. If G is a semidirect product of N by H, taking the section to be the elements of H, there is a sectional factor set such that $f(h_1, h_2) = 1$ for all $h_1, h_2 \in H$.

Example 20.7 In Example 20.5, taking x as a generator for the cyclic group $H = C_2$, the section $\{1, (1\ 2)(3\ 4\ 5)\}$ gives a sectional factor set f with values

$$f(1,1) = 1;\ f(1,x) = 1;\ f(x,1) = 1;\ \text{and}\ f(x,x) = (3\ 5\ 4);$$

whereas the section $\{1, (1\ 2)\}$, being a subgroup, produces a sectional factor set all of whose values are 1.

Remark Recall that we sometimes use exponential notation for conjugation, so that x^g denotes the product gxg^{-1}.

Proposition 20.8 *Let G be an extension of N by H and $\{s(h)\}$ be a section of G through H. The sectional factor set $f : H \times H \to N$ for the extension satisfies the following conditions:*
(i) *for all h in H, $f(1,h) = 1 = f(h,1)$;*
(ii) *for all $h_1, h_2, h_3 \in H$,*

$$f(h_1,h_2)f(h_1h_2,h_3) = f(h_2,h_3)^{s(h_1)}f(h_1,h_2h_3).$$

Proof (i) Since $s(1) = 1$,

$$f(1,h)s(h) = s(1)s(h) = s(h) = s(h)s(1) = f(h,1)s(h),$$

so that $f(1,h) = 1 = f(h,1)$.
(ii) Consider

$$\begin{aligned} s(h_1)s(h_2)s(h_3) &= (s(h_1)s(h_2))s(h_3) \\ &= f(h_1,h_2)s(h_1h_2)s(h_3) \\ &= f(h_1,h_2)f(h_1h_2,h_3)s(h_1h_2h_3). \end{aligned}$$

On the other hand,

$$\begin{aligned} s(h_1)s(h_2)s(h_3) &= s(h_1)(s(h_2)s(h_3)) \\ &= s(h_1)f(h_2,h_3)s(h_2h_3) \\ &= s(h_1)f(h_2,h_3)s(h_1)^{-1}s(h_1)s(h_2h_3) \\ &= f(h_2,h_3)^{s(h_1)}f(h_1,h_2h_3)s(h_1h_2h_3), \end{aligned}$$

so that

$$f(h_1,h_2)f(h_1h_2,h_3) = f(h_2,h_3)^{s(h_1)}f(h_1,h_2h_3),$$

as required. □

The next result is the analogue for arbitrary extensions of Proposition 19.4, recalling that for a semidirect product of N by H the elements of H can be taken as the elements of a section through H.

Proposition 20.9 *Let G be an extension of N by H with $\{s(h) : h \in H\}$ a section of G through H, and let f be the sectional factor set of the extension. For each h in H, the map $\vartheta_h : N \to N$ defined by*

$$\vartheta_h(n) = s(h)n(s(h))^{-1} = n^{s(h)}$$

is an automorphism of N. Furthermore, for all n in N and for all h_1, h_2 in H,

$$\vartheta_{h_1}\vartheta_{h_2}(n) = (\vartheta_{h_1h_2}(n))^{f(h_1,h_2)}.$$

Proof The fact that ϑ_h is an automorphism follows as in the proof of Proposition 19.4.

Now, for all $n \in N$ and $h_1, h_2 \in H$,

$$\begin{aligned}
\vartheta_{h_1}\vartheta_{h_2}(n) &= \vartheta_{h_1}(s(h_2)ns(h_2)^{-1}) \\
&= s(h_1)s(h_2)ns(h_2)^{-1}s(h_1)^{-1} \\
&= s(h_1)s(h_2)n(s(h_1)s(h_2))^{-1} \\
&= f(h_1,h_2)s(h_1h_2)n(f(h_1,h_2)s(h_1h_2))^{-1} \\
&= (s(h_1h_2)ns(h_1h_2))^{-1})^{f(h_1,h_2)} \\
&= \vartheta_{h_1h_2}(n)^{f(h_1,h_2)},
\end{aligned}$$

as required. □

Remark As well as semidirect products, other important special types of extensions are *central extensions*, which occur when the normal subgroup N is in the centre of G, and *cyclic extensions*, which occur when the group H is cyclic. Both these special cases will be considered in more detail in the next chapter.

We next consider the converse situation to the one studied so far. Thus, rather than investigating the internal structure of a known extension, we consider the reverse problem of how to build a group G out of components N and H. Since we do not have a section to work with in this context, we therefore need to start by defining a factor set in the abstract. This does not use a given extension or section, but makes use of a set of automorphisms of N instead.

Definition 20.10 *Given groups N and H, for each h in H let ϑ_h be an automorphism of N (with ϑ_1 being the trivial automorphism).* A factor set *with respect to the choice $\{\vartheta_h : h \in H\}$ is a map $f : H \times H \to N$ such that*

(i) *for all h in $H, f(1,h) = 1 = f(h,1)$; and*
(ii) *for all h_1, h_2 and h_3 in H,*

$$f(h_1,h_2)f(h_1h_2,h_3) = \vartheta_{h_1}(f(h_2,h_3))f(h_1,h_2h_3).$$

We shall say that a factor set f is compatible *if also*

(iii) *for all n in N and h_1, h_2 in H*

$$\vartheta_{h_1}\vartheta_{h_2}(n) = \vartheta_{h_1h_2}(n)^{f(h_1,h_2)}.$$

Remark Let G be an extension of N by H and $\{s(h) : h \in H\}$ be a section of G through H. By Proposition 20.9, conjugation by an element $s(h)$ of G is an automorphism ϑ_h of N (with conjugation by $s(1)$ being the trivial automorphism). Propositions 20.8 and 20.9 show that the sectional factor set for this extension is then a compatible factor set in our newly defined sense.

The next result is an analogue of Proposition 19.5 for arbitrary extensions.

Proposition 20.11 *Given groups N and H, for each h in H let ϑ_h be an automorphism of N, with ϑ_1 being the identity map on N. Suppose that $f : H \times H \to N$ is a compatible factor set. Then the set G of ordered pairs $\{(n, h) : n \in N, h \in H\}$ is a group under the multiplication*

$$(n_1, h_1)(n_2, h_2) = (n_1\vartheta_{h_1}(n_2)f(h_1, h_2), h_1h_2).$$

Furthermore, G is an extension of a group isomorphic to N by a group isomorphic to H and $\{(1, h) : h \in H\}$ is a section of G through H.

Proof We check the group axioms for G. It is clear that G is closed under the multiplication. To check associativity consider

$$\begin{aligned}
&(n_1, h_1)((n_2, h_2)(n_3, h_3)) \\
&\quad = (n_1, h_1)((n_2\vartheta_{h_2}(n_3)f(h_2, h_3), h_2h_3)) \\
&\quad = (n_1\vartheta_{h_1}(n_2\vartheta_{h_2}(n_3)f(h_2, h_3))f(h_1, h_2h_3), h_1h_2h_3) \\
&\quad = (n_1\vartheta_{h_1}(n_2)\vartheta_{h_1}\vartheta_{h_2}(n_3)\vartheta_{h_1}(f(h_2, h_3))f(h_1, h_2h_3), h_1h_2h_3).
\end{aligned}$$

Using condition (ii) of the definition of a factor set and the compatibility condition, the last line becomes

$$\begin{aligned}
&(n_1\vartheta_{h_1}(n_2)\vartheta_{h_1h_2}(n_3)^{f(h_1,h_2)}f(h_1, h_2)f(h_1h_2, h_3), h_1h_2h_3) \\
&\qquad = (n_1\vartheta_{h_1}(n_2)f(h_1, h_2)\vartheta_{h_1h_2}(n_3)f(h_1h_2, h_3), h_1h_2h_3)
\end{aligned}$$

$$= \quad ((n_1\vartheta_{h_1}(n_2)f(h_1,h_2),h_1h_2))(n_3,h_3)$$

$$= \quad ((n_1,h_1)(n_2,h_2))(n_3,h_3),$$

as required. Since $f(h,1)=1=f(1,h)$, it may easily be checked that $(1,1)$ is an identity element of G.

Finally, to show that inverses exist, notice that

$$(n,h)(\vartheta_h^{-1}(n^{-1}f(h,h^{-1})^{-1}),h^{-1}) \quad = \quad (nn^{-1}f(h,h^{-1})^{-1}f(h,h^{-1}),hh^{-1})$$
$$= \quad (1,1),$$

so that each element of G has a right inverse. As in the proof of Proposition 19.5, this is sufficient to show that

$$(\vartheta_h^{-1}(n^{-1}f(h,h^{-1})^{-1}),h^{-1})$$

is the two-sided inverse of (n,h), and that G is a group.

It remains to show that G is an extension of a group isomorphic to N by H. To do this, let N_0 be the set of elements of G of the form $\{(n,1) : n \in N\}$. Using the definition of multiplication in G,

$$(n_1,1)(n_2,1) = (n_1\vartheta_1(n_2)f(1,1),1) = (n_1n_2,1) \in N_0,$$

and also $(n,1)^{-1} = (n^{-1},1) \in N_0$, so that N_0 is a subgroup of G.

To check that N_0 is a normal subgroup of G, notice that for all $n, n_1 \in N$ and $h \in H$ there is an element g in N such that

$$(n_1,h)(n,1)(n_1,h)^{-1} = (g,h1h^{-1}) = (g,1) \in N_0.$$

It is easily checked that the map $n \mapsto (n,1)$ is an isomorphism from N to N_0. Now define a map $\vartheta : H \to G/N_0$ by $\vartheta(h) = N_0(1,h)$. This map is a homomorphism since

$$\begin{aligned}\vartheta(h_1)\vartheta(h_2) &= N_0(1,h_1)N_0(1,h_2)\\ &= N_0(1,h_1)(1,h_2)\\ &= N_0(f(h_1,h_2),h_1h_2)\\ &= N_0(1,h_1h_2)\\ &= \vartheta(h_1h_2).\end{aligned}$$

To show that ϑ is injective, if $\vartheta(h_1) = \vartheta(h_2)$, then $N_0(1,h_1) = N_0(1,h_2)$, and so N_0 contains

$$(1, h_1)(1, h_2)^{-1} = (1, h_1 h_2^{-1}).$$

This shows that $h_1 h_2^{-1} = 1$ so that $h_1 = h_2$.

The map ϑ is clearly surjective since for all n in N and h in H,

$$N_0(n, h) = N_0(n, 1)(1, h) = N_0(1, h) = \vartheta(h).$$

This completes the proof that ϑ is an isomorphism.

It only remains to show that $\{(1, h) : h \in H\}$ is a section for G through H. As we have seen, $\vartheta(h)$ is equal to $N_0(1, h)$, so that $(1, h)$ is a representative for the right coset $\vartheta(h)$. □

Corollary 20.12 *Let G be an extension of N by H, and let $\{s(h) : h \in H\}$ be a section of G through H. Let f be the corresponding sectional factor set. Then, if we define ϑ_h to be conjugation by $s(h)$, f is a compatible factor set. Conversely, given automorphisms $\{\vartheta_h : h \in H\}$ of N and a compatible factor set f, let G be the group constructed as in Proposition 20.11. Then $\{(1, h) : h \in H\}$ is a section of G through H. The map $N_0 \to N_0$ defined by $(h, 1) \mapsto (\vartheta_h, 1)$ is conjugation by $(1, h)$ and f is a sectional factor set.*

Proof The first part follows from Propositions 20.8 and 20.9. For the converse, we first consider the inner automorphism obtained by conjugating by $(1, h)$ in G. To calculate this, we use the formula for an inverse obtained during the proof of Proposition 20.11:

$$(n, h)^{-1} = (\vartheta_h^{-1}(n^{-1} f(h, h^{-1})^{-1}), h^{-1}).$$

Thus

$$\begin{aligned}
(1, h)(n, 1)(1, h)^{-1} &= (\vartheta_h(n), h)(\vartheta_h^{-1}(f(h, h^{-1})^{-1}), h^{-1}) \\
&= (\vartheta_h(n)\vartheta_h(\vartheta_h^{-1}(f(h, h^{-1})^{-1}))f(h, h^{-1}), hh^{-1}) \\
&= (\vartheta_h(n) f(h, h^{-1})^{-1} f(h, h^{-1}), 1) \\
&= (\vartheta_h(n), 1).
\end{aligned}$$

To complete the proof, note that

$$\begin{aligned}
f(h_1, h_2)(1, h_1 h_2) &= (f(h_1, h_2), h_1 h_2) \\
&= (1, h_1)(1, h_2),
\end{aligned}$$

showing that f is also a sectional factor set because $\{(1, h) : h \in H\}$ is a section of G through H. □

Remark It is not in general obvious how to find a compatible factor set when N, H and the automorphisms $\{\vartheta_h : h \in H\}$ are given. This is one reason why various restrictions are placed on the type of extensions considered. We consider two of these possibilities in the next chapter. If the map $h \mapsto \vartheta_h$ is a homomorphism, the trivial factor set is compatible, and the resulting extension is a semidirect product.

Summary for Chapter 20

The main aim of this chapter is to set up the machinery required to construct groups with given properties. The basic idea is that of an *extension* G of a group N by a group H. To understand how to find a group G which is an extension of N by H, for each $h \in H$, let ϑ_h be an automorphism of N (with ϑ_1 being the trivial automorphism). A *factor set* associated with the correspondence ϑ is a map $f : H \times H \to N$ satisfying the two conditions:

(i) for all h in $H, f(1, h) = 1 = f(h, 1)$; and

(ii) for all h_1, h_2 and h_3 in G,

$$f(h_1, h_2)f(h_1h_2, h_3) = \vartheta_{h_1}(f(h_2, h_3))f(h_1, h_2h_3).$$

There is also a compatibility condition that for all n in N and h_1, h_2 in H,

$$\vartheta_{h_1}\vartheta_{h_2}(n) = \vartheta_{h_1h_2}(n)^{f(h_1,h_2)}.$$

Then (see Proposition 20.11 for the details), the set G of ordered pairs $\{(n, h) : n \in N, h \in H\}$ with multiplication

$$(n_1, h_1)(n_2, h_2) = (n_1\vartheta_{h_1}(n_2)f(h_1, h_2), h_1h_2)$$

is an extension of N by H.

Conversely, given an extension G of N by H and a fixed choice of section $\{s(h) : h \in H\}$ of G through H, the map f defined by

$$f(h_1, h_2) = s(h_1)s(h_2)(s(h_1h_2))^{-1}$$

is a factor set satisfying the compatibility condition, where ϑ_h now denotes conjugation by $s(h)$.

The theory of group extensions was developed by Otto Hölder in the 1890s and (more famously) by Schreier in two papers in 1926.

Exercises 20

1. When G is the group $S(3)$, regarded as an extension of $A(3)$ by C_2, determine all the possible sectional factor sets of the extension.

2. Let G be the group $A(4)$ and N be the normal subgroup consisting of the four elements $\{1, (1\ 2)(3\ 4), (1\ 3)(2\ 4), (1\ 4)(2\ 3)\}$. Write down the values of the sectional factor set of G through H corresponding to the section $\{1, (1\ 2\ 3), (1\ 2\ 4)\}$. Find a section for G through H which is a subgroup of G.

3. Find all extensions of a cyclic group of order 2 by a cyclic group of order 3.

21

Central and cyclic extensions

In this chapter, we consider two particular types of extensions for which our general results can be interpreted more explicitly. The first of these occurs when each element of N is in the centre of the extension.

Definition 21.1 *A group G is a* central extension *of N by H if it is an extension of N by H and N is in the centre of G.*

Suppose that we are given such a central extension G, with section $\{s(h)\}$. Since $s(h)$ will commute with each element of N, the sectional factor set f of the extension satisfies (by Proposition 20.8) the conditions

(i) for all h in $H, f(1,h) = 1 = f(h,1)$; and
(ii) for all h_1, h_2, h_3 in H, $f(h_1,h_2)f(h_1h_2,h_3) = f(h_2,h_3)f(h_1,h_2h_3)$.

We shall refer to a factor set satisfying these two conditions as a *central factor set.*

Example 21.2 The quaternion group with 8 elements, $G = Q_8$, has presentation

$$\langle x, y : x^2 = y^2,\ x^4 = 1,\ x^{-1}yx = y^{-1}\rangle.$$

The element $z = x^2$ has order 2. This element z, being a power of x, commutes with all powers of x, and also commutes with every power of y since z is also a power of y. It follows that z therefore commutes with each element of $G = \langle x, y\rangle$, so that $z \in Z(G)$. Since G is non-abelian, $Z(G)$ cannot have more than two elements by Proposition 10.21, so $Z(G)$ is the subgroup generated by z. The quotient group $G/Z(G)$ has presentation

$$\langle \bar{x}, \bar{y} : \bar{x}^2 = \bar{y}^2 = \bar{1},\ \bar{y}\bar{x} = \bar{x}\bar{y}\rangle,$$

where $\bar{g}$ denotes the right coset $Z(G)g$, and so $G/Z(G)$ is isomorphic to the direct product $C_2 \times C_2$. Thus G is a central extension of C_2 by H, the group $\langle a, b : a^2 = b^2 = 1,\ ab = ba\rangle$. We now determine the values of a sectional factor set for G through H. First choose a section of G through H by $s(1) = 1$, $s(a) = x$, $s(b) = y$ and $s(ab) = xy$. Then, for example,

$$s(b)s(a)s(ab)^{-1} = yx(xy)^{-1} = x^{-1}yy^{-1}x^{-1} = x^{-2} = x^2 = z.$$

This calculation can be repeated for other group elements to see that the factor set f associated with this section has the following values:

f	1	a	b	ab
1	1	1	1	1
a	1	z	1	z
b	1	z	z	1
ab	1	1	z	z

Since z is in the centre of G, it follows by Proposition 20.8 that f is a central factor set and also that f satisfies the compatibility condition.

Suppose now that we wish to construct a central extension of N by H. To do this, we choose each automorphism $\{\vartheta_h : h \in H\}$ to be the identity automorphism of N. The compatibility condition then holds for trivial reasons and the factor set conditions become the requirements that f be a central factor set:

(i) for all h in H, $f(1,h) = 1 = f(h,1)$;
(ii) for all h_1, h_2, h_3 in H, $f(h_1,h_2)f(h_1h_2,h_3) = f(h_2,h_3)f(h_1,h_2h_3)$.

Example 21.3 Consider the central extension G of a cyclic group N of order 2, generated by z, by the group H which is an internal direct product $C_2 \times C_2$, generated by a and b. We have already seen a central factor set f for this extension in Example 21.2 above, and we use that to produce another factor set g using a method which generalises to other situations. Let w be an element of order 4 such that $w^2 = z$ (formally, w is an element of the central product $N \times_\vartheta C_4$ as defined in Chapter 13). Define a map δ from H to $\langle w \rangle$ with values given by

$$\delta(1) = 1 = \delta(a); \; \delta(b) = w^{-1}; \text{ and } \delta(xy) = w.$$

Then define F by the equation

$$F(x,y) = \delta(x)\delta(y)\delta(xy)^{-1}f(x,y)$$

for all x, y in H. It is then straightforward to calculate the following values for F:

F	1	a	b	ab
1	1	1	1	1
a	1	z	z	1
b	1	1	1	1
ab	1	z	z	1

The fact that F is a central factor set can be seen as follows. For condition (i): for all $h \in H$,

$$F(1,h) = \delta(1)\delta(h)\delta(h)^{-1}f(1,h) = 1 = f(h,1) = F(h,1).$$

To check (ii), note that the values of δ all lie in $\langle w \rangle$, and so these commute with z, since $w^2 = z$. Thus

$$\begin{aligned} F(x,y)F(xy,z) &= \delta(x)\delta(y)\delta(xy)^{-1}\delta(xy)\delta(z)\delta(xyz)^{-1}f(x,y)f(xy,z) \\ &= \delta(x)\delta(y)\delta(z)\delta(xyz)^{-1}f(x,y)f(xy,z) \\ &= \delta(x)\delta(y)\delta(z)\delta(xyz)^{-1}f(y,z)f(x,yz) \\ &= F(y,z)F(x,yz). \end{aligned}$$

Since the values of F are known, we may construct the corresponding group using the method in the proof of Proposition 20.11, taking each ϑ_h for $h \in H$ to be the trivial automorphism. We therefore define $(1,a)$ to be x, so that

$$x^2 = (1,a)^2 = (1,a^2)F(a,a) = z.$$

Define $(1,b)$ to be y, so that

$$y^2 = (1,b^2)F(b,b) = 1.$$

Also x has order 4, so $x^{-1} = x^3$ and

$$\begin{aligned} yxy &= (1,b)(1,a)(1,b) = F(b,a)(1,ba)(1,b) \\ &= F(ba,b)(1,bab) = F(ab,b)(1,a) \\ &= z(1,a) = x^2x = x^{-1}. \end{aligned}$$

Thus G is non-abelian of order 8 and has elements x, y satifying

$$\langle x, y : x^4 = y^2 = 1,\ yxy = x^{-1} \rangle.$$

It follows that G is isomorphic to the dihedral group of order 8.

Remark We have seen that the dihedral group $D(4)$ and the quaternion group Q_8 are both central extensions of C_2 by $C_2 \times C_2$. Other important examples of central extension are the *double covers* of the symmetric groups. These are groups of order $2n!$. A concrete realisation of one of these double covers can be given for $S(4)$. Consider the group G of all 2×2 invertible matrices over $\mathbf{Z}_3$. There are eight ways to chose the first column of such a matrix (any non-zero column vector), and then six ways to choose the second (any vector which is not a scalar multiple of the first column). Thus G has 48 elements. The matrix $2I_2$ is in the centre of the group, and so the subgroup $Z = \{I_2, 2I_2\}$ is a normal subgroup of G. Now let

$$A_1 = \begin{pmatrix} 2 & 0 \\ 0 & 1 \end{pmatrix}, \; A_2 = \begin{pmatrix} 2 & 2 \\ 0 & 1 \end{pmatrix}, \; A_3 = \begin{pmatrix} 0 & 1 \\ 1 & 0 \end{pmatrix}.$$

It is easy to check that

$$A_1^2 = A_2^2 = A_3^2 = I_2, (A_1A_2)^3 = I_2 = (A_2A_3)^3, \text{ and } (A_1A_3)^2 = 2I_2.$$

It follow from this that the cosets A_1Z, A_2Z and A_3Z generate a group isomorphic to $S(4)$, so that the group G is a double cover of $S(4)$.

Definition 21.4 *The second special type of extension which we consider is that of a cyclic extension of N by H. This occurs when H is a cyclic group.*

In the case of cyclic extensions, it is possible to give an explicit description which enables one to determine the extensions of N by H.

Theorem 21.5 *Let G be an extension of a group N by a cyclic group $H = \langle h \rangle$ of order r. Choose $s(h)$ to be an element of G such that $Ns(h)$ is a generator for G/N, so that $(s(h))^r = n_0$ for some $n_0 \in N$. Then there is an automorphism ϑ of N satisfying the conditions:*

(a) *ϑ^r is conjugation by the element n_0 of N;*
(b) *ϑ fixes n_0.*

Conversely, given an element n_0 of N and an automorphism ϑ of N satisfying (a) *and* (b), *the set of ordered pairs (n, h^i) with $n \in N$ and $0 \leq i \leq r-1$ is a cyclic extension of N by H under the multiplication*

$$(n_1, h^i)(n_2, h^j) = (n_1(\vartheta)^i(n_2)f(h^i, h^j), h^{i+j})$$

where

$$f(h^u, h^v) = \begin{cases} 1 & \text{if } u+v < r \\ n_0 & \text{if } u+v \geq r. \end{cases}$$

It follows from this that $(1, h)^r = (n_0, 1)$.

Proof Let ϑ be the automorphism of N given by conjugation by $s(h)$. Since n_0 is a power of $s(h)$, conjugation of n_0 by $s(h)$ gives n_0 and so ϑ fixes n_0. Furthermore, ϑ^r is conjugation by $(s(h))^r = n_0$. This shows that conditions (a) and (b) hold.

Conversely, in order to apply the construction of Proposition 20.11, we must show that the given f is a compatible factor set. It will then be clear that G is an extension of N by the cyclic group H.

The factor set condition (i) of Definition 20.10 is clear from the definition of f since if either u or v is zero, then $f(h^u, h^v) = 1$.

The fact that ϑ fixes n_0 means that each ϑ^i fixes each value of f. Thus, if we set $\vartheta_{h^i} = \vartheta^i$ for $1 \leq i \leq r-1$, the factor set condition (ii) becomes

$$f(h^u, h^v)f(h^{u+v}, h^w) = f(h^v, h^w)f(h^u, h^{v+w})$$

for all h^u, h^v and h^w in H. To see why this equation holds, we consider three possibilities:

(1) If $u + v + w < r$, then all terms on both sides of (ii) are equal to 1.

(2) If $r \leq u + v + w < 2r$, then precisely one term on each side of condition (ii) is equal to n_0. Either

$u+v$ is less than r, in which case $f(h^u, h^v) = 1$ and $f(h^{u+v}, h^w) = n_0$,

or

$u + v$ is greater than r, so that $f(h^u, h^v) = n_0$, but since h^{u+v} is then equal to h^s for some $s \leq r$, then $f(h^{u+v}, h^w) = 1$.

A similar analysis works for the right-hand side of (ii).

(3) If $2r \leq u + v + w < 3r$, then we can show, as in (2), that each of $f(h^u, h^v)$, $f(h^{u+v}, h^w)$, $f(h^v, h^w)$ and $f(h^u, h^{v+w})$ must equal n_0.

Finally, to prove that the compatibility condition of Definition 20.10 holds, notice that if $u + v < r$, then $f(h^u, h^v) = 1$, and so

$$\vartheta_{h^u}\vartheta_{h^v} = \vartheta_{h^{u+v}}.$$

If, however, $u+v = r+w$ with $w \geq 0$, then $f(h^u, h^v) = n_0$ and $h^u h^v = h^w$, so that for all $n \in N$,

$$\begin{aligned}
\vartheta_{h^u}\vartheta_{h^v}(n) &= \vartheta^u\vartheta^v(n) \\
&= \vartheta^{r+w}(n) \\
&= (\vartheta^w(n))^{n_0} \\
&= (\vartheta_{h^{u+v}})^{f(h_u, h_v)},
\end{aligned}$$

as required. The result then follows by Proposition 20.11. Note that an easy induction proof shows that $(1, h)^i = (1, h^i)$ for $1 \leq i \leq r-1$. Also

$$(1,h)^r = (1,h)(1,h)^{r-1} = (1,h)(1,h^{r-1}) = (1(\vartheta)(1)f(h,h^i),h^r) = (n_0,1).$$

Thus, if N_0 denotes the subgroup of elements of the form $\{(n,1) : n \in N\}$, the quotient group G/N_0 generated by the coset of $(1,h)$ is cyclic of order r. □

Corollary 21.6 *Let G be an extension of an abelian group N by a cyclic group H of order r and let $Ns(h)$ be a generator for G/N. Then there is an automorphism ϑ of N with ϑ^r the identity automorphism, and an element n_0 of N fixed by ϑ with $n_0 = (s(h))^r$. Conversely, in order to construct a cyclic extension of N by H we need to be given an automorphism ϑ of N of order r and an element of N fixed by ϑ. The group G then consists of elements $ns(h^i)$, where $s(h)^r$ is the chosen element of N fixed by ϑ.*

Proof The result follows from Theorem 21.5 since any inner automorphism of an abelian group is trivial. □

Example 21.7 We return to the classification of groups of order 8 which was considered in Chapter 5 and re-establish this result using extension theory. Let G be a group of order 8. If G has an element of order 8, then G is cyclic. If each element of G has order 2, then G is abelian by Exercise 3.1, and so is isomorphic to $C_2 \times C_2 \times C_2$. We may therefore suppose that G is not cyclic and that G has an element x, say, of order 4, so that $\langle x \rangle$, being of index 2, is normal. Thus G is an extension of a cyclic group of order 4 by a cyclic group of order 2. Let y be any element of G not in $\langle x \rangle$. In order to construct G, we therefore need to find an element n_0 of $N = \langle x \rangle$ and an automorphism ϑ fixing n_0 such that ϑ^2 is the identity automorphism (by Corollary 21.6). An automorphism of $\langle x \rangle$ is specified by its value on x, and since this value is an element of order 4, the only possible automorphisms are the identity map and the map taking x to x^{-1}. In the case that $\vartheta(x) = x$, the subgroup $\langle x \rangle$ is central and so G is abelian, being a cyclic extension of a central subgroup. It then follows by the Classification Theorem for finite abelian groups that G is isomorphic to $C_4 \times C_2$. In the case that $\vartheta(x) = x^{-1}$, the element $y^2 = n_0$ is a fixed point of the automorphism ϑ, so y^2 is either 1 or x^2. In the first case G is isomorphic to the dihedral group $D(4)$, and in the second case G is isomorphic to the quaternion group Q_8 of order 8.

Remark On several occasions in earlier chapters, we have ended a discussion on the classification of groups of various orders with a list of possible presentations. Until now, we have not had a method to ensure that a given presentation does not collapse, in that the relations could lead to unexpected further relations which reduce the order of the group. We have often dealt with this problem by giving explicit 'models' of the group as permutations or matrices. The results of extension theory provide a general method to deal with these existence problems. For example, we concluded the discussion of groups of order p^3, for p an odd prime, by deducing that a non-abelian group of this order is isomorphic to one of

$$\langle x, y : x^{p^2} = 1 = y^p,\ yxy^{-1} = x^{1+p}\rangle \text{ or}$$

$$\langle x, y, z : x^p = 1 = y^p = z^p,\ yx = zxy,\ xz = xz,\ yz = zy\rangle.$$

To see why these groups exist, note that the first is a cyclic extension of a cyclic group of order p^2, with automorphism $x \mapsto x^{1+p}$ of order p and with $n_0 = 1$. In the second case, the normal subgroup is $\langle x, z\rangle$ (which is isomorphic to $C_p \times C_p$), the automorphism ϑ is determined by $\vartheta(x) = zx$ and $\vartheta(z) = z$, and here again n_0 is 1.

We have, finally, determined the isomorphism types of groups of order p^3 when p is odd. It is now clear that, apart from the cyclic group, each group of order p^3 has a subgroup of the form $C_p \times C_p$. The corresponding result for $p = 2$ fails because the quaternion group Q_8 has only one element of order 2. In general, we have the following fact:

Proposition 21.8 *Let p be an odd prime and let G be a finite p-group which has only one subgroup of order p. Then G is cyclic.*

Proof The proof is by induction on $|G| = p^n$. The result holds trivially for groups with p or p^2 elements. We therefore suppose that $n \geq 3$ and that the result holds for groups of order p^m with $m < n$. Any subgroup of G can have only one subgroup of order p and so is cyclic by induction. In particular, if N is a maximal subgroup of G, then N is cyclic and also normal by Corollary 18.7. The maximality of N means that G/N is cyclic of order p. Let x be a generator for N, so that x has order p^{n-1}. Since $n \geq 3$, the subgroup K generated by $x^{p^{n-3}}$ is a non-trivial characteristic subgroup of N by Example 16.7. It then follows by Proposition 16.8 that K is a normal subgroup of G. The quotient group G/K is therefore of order p^3. The inductive hypothesis implies that this quotient group cannot have s subgroup isomorphic to $C_p \times C_p$, since such a subgroup would force G to have a non-cyclic subgroup of index p. We now use the list of isomorphism types of groups of order p^3 given in Proposition 18.11 to deduce that G/K is cyclic.

If g is an element of G with gK a generator for G/K, we know that g^pK generates N/K, so that $g^pK = x^iK$ for some i coprime to p^2. Thus $g^p = x^{i+jp^{n-3}}$ for some j. This means that g^p generates N so that G is cyclic, as required. □

Remark It may be shown that a 2-group which only has one element of order 2 is either cyclic or a generalised quaternion group.

To conclude the chapter, we give a simple application of Corollary 21.6.

Proposition 21.9 *Let p and q be prime integers with $p > q$. A group G of order pq has presentation*

$$\langle x, y : x^p = 1 = y^q, yxy^{-1} = x^k \text{ with } k^q \equiv 1 \text{ mod } p \rangle.$$

If q does not divide $p-1$, then a group of order pq is cyclic.

Proof Since G has order pq, Proposition 12.1 shows that G must have one Sylow p-subgroup, so that this subgroup, P, is normal. It follows that G is an extension of a cyclic group of order p by a cyclic group of order q. An automorphism of a cyclic group of order p is determined by its value on a generator x. Since there are are $p-1$ possible choices for this value, $|\text{Aut}(C_p)| = p-1$. If x is a generator for P and y is a generator for a Sylow q-subgroup Q, we see that $x^p = 1 = y^q$. Since P is a normal subgroup of G, $yxy^{-1} = x^k$ for some k. Since conjugation by y is an automorphism of order dividing q, it follows that $k^q \cong 1$ mod p, as required. If q does not divide $p-1$, then Sylow theory implies that G also has a unique Sylow q-subgroup, Q. It then follows that G is an internal direct product of P and Q, so that G is isomorphic to $C_p \times C_q$, and so G is cyclic by Proposition 13.1(iii). □

Summary for Chapter 21

There are two special types of extension which are investigated in this chapter. An extension is *central* if each element of N is in the centre of the extension. In order to determine an extension in this case, take all the automorphisms ϑ_h to be the trivial automorphisms. The requirements for a compatible factor set then simplify to:

(i) for all h in $H, f(h,1) = 1 = f(1,h)$; and
(ii) $f(h_1,h_2)f(h_1h_2,h_3) = f(h_2,h_3)f(h_1,h_2h_3)$ for all h_1, h_2 and h_3 in H.

The other special case is that of a *cyclic* extension, when the group H is cyclic. This is a situation where it is possible to give a more precise description of how to construct an extension from N and H. In fact, to construct an extension of N by the cyclic group $H = \langle h \rangle$ of order r, we need to be given an element n_0 of N and an automorphism ϑ of N satisfying

(a) ϑ^r is conjugation by the element n_0 of N; and
(b) ϑ fixes n_0.

The group G then consists of elements $ns(h^i)$ where $s(h)^r = n_0$.

One reason for the importance of cyclic extensions is that finite soluble groups are built up from cyclic extensions, and so this enables one in principle to construct all finite soluble groups.

The theory of central extensions was developed by Schur at the turn of the century and later by Schreier. The theory of cyclic extensions was first discussed by Hölder and later Schreier.

Exercises 21

1. Let $N = \langle z \rangle$ be a cyclic group of order 2, and H be the non-cyclic group of order 4 with elements $\{1, a, b, c\}$. Given that the factor set

f	1	a	b	ab
1	1	1	1	1
a	1	1	z	z
b	1	1	z	z
ab	1	1	1	1

is a central factor set, construct the central extension G. Is G isomorphic to $D(4)$?

2. Determine all the central extensions of C_3 by the symmetric group $S(3)$.

3. Find all extensions of an infinite cyclic group by a group of order 2.

22

Groups with at most 31 elements

In this chapter, we shall apply the results established in the earlier chapters to produce examples of groups with various properties. One of the main objectives is to determine the complete list of groups with at most 31 elements. Since we shall need to know more about automorphism groups, we first calculate the automorphism groups of some standard groups.

Proposition 22.1 *Let p be a prime integer. The automorphism group of a cyclic group of order p is cyclic of order $(p-1)$.*

Proof Let $G = \langle x : x^p = 1 \rangle$. An automorphism ϑ of G is completely determined by knowing the value of ϑ on x. Since the only non-identity divisor of p is p itself, it follows by Lagrange's Theorem that, if i is not divisible by p, x^i is an element of G of order p. Thus there are $p-1$ possible values of i such that x^i has order p. Since each of these choices $x \mapsto x^i$ determines an automorphism $x^k \mapsto x^{ki}$ of G, it follows that $\mathrm{Aut}(G)$ has order $p-1$. Since the maps $x^k \mapsto x^{ki}$ and $x^k \mapsto x^{kj}$ commute, it is clear that $\mathrm{Aut}(G)$ is abelian. If $\mathrm{Aut}(G)$ is not cyclic, there is a prime q such that the Sylow q-subgroup of $\mathrm{Aut}(G)$ is non-cyclic. It follows that this Sylow q-subgroup has at least two generators. Thus, by Proposition 14.9, this Sylow q-subgroup contains a subgroup isomorphic to $C_q \times C_q$. Each element of $\mathrm{Aut}(G)$ of order q corresponds to an integer i in $\mathbf{Z}_p$ such that $x^{i^q} = x$. Thus there are at least q^2 elements i in $\mathbf{Z}_p$ such that $i^q = 1$. In that case there are q^2 roots of the polynomial $X^q - 1$ over the field $\mathbf{Z}_p$. This contradicts the elementary theory of polynomials over fields, in that a polynomial of degree q can have at most q roots. It follows that the group $\mathrm{Aut}(G)$ is cyclic. □

Remark We have seen this result before in a slightly different guise. It was shown in Proposition 14.15 that the multiplicative group of a finite field is cyclic. Applying this to $\mathbf{Z}_p$ shows that the multiplicative group of $\mathbf{Z}_p$ is cyclic of order $p-1$. As we have seen, the elements in $\mathrm{Aut}(G)$ are in bijective correspondence with non-zero elements of $\mathbf{Z}_p$. This correspondence is also a homomorphism and hence $\mathrm{Aut}(G)$ is cyclic.

Example 22.2 Consider the case when $p = 17$, so that (by Proposition 22.1) there is an element $i \in \mathbf{Z}_{17}$ with $i^{16} \equiv 1 \bmod 17$, but with no smaller power of i congruent to 1 modulo 17. It needs trial and error to see that 3 is one of the generators for the multiplicative group of $\mathbf{Z}_{17}$ since

$$3^2 \equiv 9 \mod 17;\ 3^4 \equiv (-4) \mod 17; \text{and } 3^8 \equiv -1 \mod 17.$$

Hence the automorphism $g \mapsto g^3$ is of order 16, and therefore generates $\mathrm{Aut}(C_{17})$. In general, there is no systematic way to determine an element which generates the multiplicative group of $\mathbf{Z}_p$.

Proposition 22.3 *For $n > 2$, the automorphism group of a cyclic group G of order 2^n is a direct product $C_2 \times C_k$, where $k = 2^{n-2}$.*

Proof Let x be a generator for the group G. An automorphism ϑ of G is specified by the value $\vartheta(x)$. In fact x^i has order 2^n if and only if i is odd, so $\mathrm{Aut}(G)$ has 2^{n-1} elements. The automorphism $x \mapsto x^5$ has order 2^{n-2} since

$$5^{2^{n-2}} \equiv (1+4)^{2^{n-2}} \equiv 1 + 2^n + \cdots \equiv 1 \mod 2^n,$$

whereas

$$5^{2^{n-3}} \equiv 1 + 2^{n-1} \mod 2^n$$

Let H be the subgroup generated by the automorphism $x \mapsto x^{-1}$. Then H is a subgroup of $\mathrm{Aut}(G)$ which is normal because $\mathrm{Aut}(G)$ is abelian. Since no power of 5 is congruent to -1 modulo 2^n, it follows by Proposition 13.5 that $\mathrm{Aut}(G)$ is the internal direct product $H \times K$, where K is the subgroup generated by the automorphism $x \mapsto x^5$. □

Remark If p is odd, the automorphism group of a cyclic group of order p^n has order $p^{n-1}(p-1)$, since there are $(p^n - p^{n-1})$ possible values of i which ensure that x^i has order p^n. Furthermore, it can be shown that this automorphism group is itself cyclic. The proof of this fact will not be needed for our purposes. In contrast to the case when p is 2, the proof that $\mathrm{Aut}(G)$ is cyclic is not constructive. Thus the proof shows that a generator for $\mathrm{Aut}(G)$ exists, but does not tell us what this generator is in the case where the prime is odd. In small cases it is possible to see the result directly. For example, the automorphism group of a group of order 9 has order 6, and a little experimentation shows that the map taking x to x^2 has 6 distinct powers.

Proposition 22.4 *Let p be a prime and G be an elementary abelian p-group of order p^n, so that G is a direct product of n copies of C_p. The automorphism group of G is isomorphic to the group $GL(n,p)$ of invertible $n \times n$ matrices with entries in the finite field $\mathbf{Z}_p$.*

Proof The case when $n = 1$ is established in Proposition 22.1 since an invertible 1×1 matrix over $\mathbf{Z}_p$ can be identified with its unique (non-zero) entry. Choose a set $x_1, x_2, \ldots, x_n$ of generators for G. Each automorphism ϑ of G is uniquely determined once $\vartheta(x_i)$ (for $1 \leq i \leq n$) is known, since

$$\vartheta(x_1^{a_1} x_2^{a_2} \ldots x_n^{a_n}) = \vartheta(x_1)^{a_1} \vartheta(x_2)^{a_2} \ldots \vartheta(x_n)^{a_n}.$$

However, for each x in G, the element $\vartheta(x)$ is of the form

$$x_1^{\alpha_1} x_2^{\alpha_2} \ldots x_n^{\alpha_n},$$

with

$$\alpha_1, \alpha_2, \ldots, \alpha_n \in \mathbf{Z}_p.$$

It is clear therefore that we may represent ϑ as an $n \times n$ matrix over $\mathbf{Z}_p$ in which the ith column records the powers of the generators occurring in $\vartheta(x_i)$. The fact that ϑ is an automorphism then translates into the fact that the corresponding matrix is invertible. This correspondence between automorphisms and invertible matrices is injective (since distinct automorphisms correspond to distinct matrices) and is surjective (because each invertible matrix uniquely determines an automorphism). Finally, the correspondence is a homomorphism since multiplication of matrices corresponds to composition of maps. □

Example 22.5 The automorphism group of $C_2 \times C_2$ is, by Proposition 22.4, isomorphic to the group $GL(2, 2)$ of invertible 2×2 matrices over $\mathbf{Z}_2$. There are three choices for the first column of such a matrix since the column with both entries equal to zero would produce a matrix of determinant zero. The second column must also be non-zero and also be different from the first column in order to ensure that the determinant is non-zero. There are therefore two choices for the second column. Thus there are six invertible 2×2 matrices over Z_2, namely

$$\begin{pmatrix} 1 & 0 \\ 0 & 1 \end{pmatrix}, \begin{pmatrix} 1 & 1 \\ 0 & 1 \end{pmatrix}, \begin{pmatrix} 0 & 1 \\ 1 & 0 \end{pmatrix}, \begin{pmatrix} 0 & 1 \\ 1 & 1 \end{pmatrix}, \begin{pmatrix} 1 & 1 \\ 1 & 0 \end{pmatrix} \text{ and } \begin{pmatrix} 1 & 0 \\ 1 & 1 \end{pmatrix}.$$

To interpret each of these matrices as automorphisms of the internal direct product $G = C_2 \times C_2$, let a and b be elements of order 2 generating G (corresponding to x_1 and x_2 in the proof of Proposition 22.4). The automorphism ϑ corresponding to the second matrix above would therefore have $\vartheta(a) = a$ and $\vartheta(b) = ab$. Since ϑ is an automorphism, it would follow that

$$\vartheta(ab) = \vartheta(a)\vartheta(b) = aab = b.$$

Proposition 22.6 *The automorphism group of the dihedral group $D(3)$ is isomorphic to $D(3)$.*

Proof To see this, let b be an element of order 2 in $D(3)$ and a be an element of order 3. Since the centre of $D(3)$ is trivial, it follows by Proposition 8.17 that the inner automorphism group of $D(3)$ is isomorphic to $D(3)$ itself. In particular, $\text{Inn}(D(3))$ has six elements. We saw in Exercise 8.4 that any automorphism maps an element g to one with the same order as that of g. It follows that an automorphism of $D(3)$ has two possible images for a, and three possible images for b. There are therefore at most 6 automorphisms, and since there are six inner automorphisms, every automorphism is inner, and so $\text{Aut}(D(3)) \cong D(3)$. □

We now start our investigations of groups of given orders by gathering together some of the results from earlier sections. In the following, p and q always denote prime integers.

(1) A group of order p is cyclic (Proposition 5.19).

(2) A group of order p^2 is abelian and is isomorphic either to C_{p^2} or to $C_p \times C_p$ (Corollaries 10.22 and 10.23).

(3) An abelian group of order p^3 is isomorphic to one of C_{p^3}, $C_{p^2} \times C_p$ or $C_p \times C_p \times C_p$. A non-abelian group of order p^3 is isomorphic to the dihedral or quaternion group when $p = 2$, and for odd p, the group is either

$$\langle x, y, z : x^p = y^p = z^p = 1,\ xz = zx,\ yz = zy,\ x^{-1}y^{-1}xy = z\rangle$$

in which every non-identity element has order p, or

$$\langle x, y : x^{p^2} = y^p = 1,\ y^{-1}xy = x^{1+p}\rangle,$$

which has an element of order p^2 (Proposition 18.11).

(4) A group of order pq, with $p > q$, is cyclic unless q divides $p - 1$, in which case there is also a non-abelian group pq with presentation

$$\langle x, y : x^p = 1 = y^q,\ y^{-1}xy = x^k, \text{ with } k \not\equiv 1 \bmod p \text{ and } k^q \equiv 1 \bmod p\rangle$$

(Proposition 21.9).

It is convenient to recall some of our standard classes of groups. As well as cyclic groups, which exist for every possible order, for every even integer $2n$, there is a dihedral group $D(n)$ of order $2n$ with presentation

$$\langle a, b : a^n = 1 = b^2,\ bab^{-1} = a^{-1}\rangle.$$

As we have seen in Chapter 17, the commutator quotient group of $D(n)$ is C_2 if n is odd, and is $C_2 \times C_2$ if n is even. There is another standard group for integers of the form $4n$.

Proposition 22.7 *For any integer n, there is a group of order $4n$ with generators a and b satisfying the relations*

$$a^{2n} = 1, \ a^n = b^2, \ bab^{-1} = a^{-1}.$$

This is the generalised quaternion group Q_{4n}. Its derived group is generated by a^2, and its centre is generated by a^n.

Proof The existence of Q_{4n} follows from our general methods. The group is an extension of a cyclic group of order $2n$ by a cyclic group of order 2. Since a^n is a fixed point of the automorphism determined by $x \mapsto x^{-1}$, we may set $b^2 = a^n$. The claims concerning the centre and derived group are easily checked. □

The intention now is to consider the list of positive integers up to 31 with the aim of producing the isomorphism types of finite groups with each possible order. We shall see that the difficulty in this project does not usually lie in finding groups of order n, but rather in showing that the list is complete. The first integer not already dealt with is 12.

Proposition 22.8 *There are five isomorphism classes of groups with 12, namely C_{12}, $C_6 \times C_2$, $D(6)$, the alternating group $A(4)$, and the generalised quaternion group Q_{12} with presentation*

$$\langle a, b : a^6 = 1, \ a^3 = b^2, \ bab^{-1} = a^{-1} \rangle.$$

Proof We saw in Proposition 12.3, as a consequence of Sylow's Theorems, that any group G with 12 elements either has one Sylow 2-subgroup or has one Sylow 3-subgroup. We consider these possibilities separately.

Case 1. Suppose that the group G has a unique Sylow 2-subgroup T, say. Since T has order 4, it could either be cyclic or isomorphic to $C_2 \times C_2$. In either case, G is an extension of a group of order 4 by a cyclic group (of order 3). Suppose first that T is cyclic generated by x, and let y be an element of order 3 (so that y is a generator of a Sylow 3-subgroup of G). The automorphism group of a cyclic group of order 4 is of order 2, since a generator is either fixed or inverted. Since the automorphism induced on T by conjugation by y has order dividing 3, it follows that the $yxy^{-1} = x$ and so G is abelian. Thus G is, in fact, cyclic of order 12 in this case.

We therefore suppose that every non-identity element of T has order 2. By Proposition 22.4, the automorphism group of T is isomorphic to the group $GL(2,2)$. The only elements of $GL(2,2)$ of order 3 are

$$\begin{pmatrix} 0 & 1 \\ 1 & 1 \end{pmatrix} \text{ and its inverse } \begin{pmatrix} 1 & 1 \\ 1 & 0 \end{pmatrix}.$$

Thus $GL(2,2)$ has a unique subgroup of order 3. If the group G is abelian, since its Sylow 2-subgroup is non-cyclic, G is $C_6 \times C_2$. Using the theory of cyclic extensions, it follows from Corollary 21.6 that if G is non-abelian, it has presentation

$$\langle a, b, c : \ a^2 = b^2 = c^3 = 1, \ ab = ba, \ cac^{-1} = b, \ cbc^{-1} = ab \rangle.$$

The other possible value for the automorphism (with $cac^{-1} = ab$ and $cbc^{-1} = a$) gives an isomorphic group (since c^2 conjugates a to b and b to ab).

The group G with the above presentation is isomorphic to the alternating group $A(4)$ by the map

$$a \mapsto (1\ 2)(3\ 4); \ b \mapsto (1\ 4)(2\ 3); \ c \mapsto (1\ 2\ 3),$$

since

$$(1\ 2\ 3)(1\ 2)(3\ 4)(1\ 2\ 3)^{-1} = (1\ 4)(2\ 3); \text{ and}$$
$$(1\ 2\ 3)(1\ 4)(2\ 3)(1\ 2\ 3)^{-1} = (1\ 3)(2\ 4) = (1\ 2)(3\ 4)(1\ 4)(2\ 3),$$

and hence the group generated by a, b and c is isomorphic to $A(4)$.

We have seen that in the case in which the Sylow 2-subgroup is normal, the group G is isomorphic to C_{12}, $C_6 \times C_2$ or $A(4)$.

Case 2. Now suppose that the Sylow 3-subgroup P of G is normal. The quotient group G/P is then of order 4 and so is abelian. It follows that G/P has a subgroup of order 2 and so G has a normal subgroup N, say, of order 6 and index 2. Suppose, in the first instance, that this subgroup N is cyclic, say $N = \langle x \rangle$. Let y be any element of G not in N, so that y^2 is in N. Then by Corollary 21.6, the automorphism ϑ_y of N has order dividing 2. It follows that yxy^{-1} is either x or x^{-1}. In the first case, G is abelian and so is either C_{12} or $C_6 \times C_2$. In the second, y^2 is a fixed point of the inner automorphism induced by conjugation by y, so is either 1 or x^3. The corresponding groups are $D(6)$ and the generalised quaternion group

$$\langle x, y : x^6 = 1, \ x^3 = y^2, \ yxy = x^{-1} \rangle.$$

It only remains to show that in the case when the normal subgroup N is isomorphic to the dihedral group $D(3)$, no new groups appear, so that the list of groups with 12 elements is complete. We show that if N is isomorphic to $D(3)$, the group G of order 12 must also have a cyclic normal subgroup of order 6, so that the group is covered by our previous discussion.

The first thing to note is that, by Proposition 22.6, the automorphism group of $D(3)$ is $D(3)$, so that every automorphism of $D(3)$ is inner. Let N be isomorphic to $D(3)$ and generated by an element a of order 3 and an element b of order 2. Let y be an element of G not in N. It follows by Corollary 21.6 that y^2 is equal to some element z of N. Hence, the square of the inner automorphism induced by conjugation by y is the inner automorphism induced by the element z, and also $yzy^{-1} = z$. If y centralises a, conjugation by y is the inner automorphism obtained either from the identity element, from a or from a^2. It follows that y^2 is either $1, a$ or a^2, so that $\langle a, y\rangle$ is a cyclic subgroup of order 6 and index 2. This means that we return to the previous situation. If y does not centralise a, then conjugation by y inverts a. In this case conjugation by y is the inner automorphism associated with b, ba or ba^2. In this case, by centralises a. If by has order 4, then, since G has 12 elements, G is generated by a and by. Since G is generated by commuting elements, it is therefore abelian. This gives a contradiction to the fact that G has a non-abelian normal subgroup. We therefore conclude that by has order 2, so that $\langle a, by\rangle$ is a cyclic normal subgroup of order 6, returning us to the previous case. □

The methods we have outlined have already enabled us to determine the isomorphism classes of groups with 13, 14 or 15 elements. The next case which requires some discussion is that of groups with 16 elements. Since any 2-group is nilpotent, a maximal subgroup of a group of order 16 has 8 elements and is normal. We can then use the method of Theorem 21.5 to complete the list of isomorphism types. These are given in Appendix B, but one of the cases which needs to be considered (the case when there is a cyclic subgroup with 8 elements), is easily generalised to give the following result.

Proposition 22.9 *Let n be an integer greater than 2. Let G be a group of order 2^{n+1} which has an element of order 2^n. If G is abelian, G is isomorphic either to a cyclic group or to the group $C_{2^n} \times C_2$. If G is non-abelian, G is isomorphic to one of:*

(a) *the dihedral group $D(2^n)$;*

(b) *the generalised quaternion group, $Q_{2^{n+1}}$, with presentation*

$$\langle a, b : a^{2^n} = 1,\ b^2 = a^{2^{n-1}},\ bab^{-1} = a^{-1}\rangle,$$

(c) *the group*

$$\langle x, y : x^{2^n} = y^2 = 1,\ y^{-1}xy = x^{2^{n-1}+1}\rangle, \text{ or}$$

(d) *the quasi-dihedral group*

$$\langle x, y : x^{2^n} = y^2 = 1,\ y^{-1}xy = x^{2^{n-1}-1}\rangle.$$

Proof Apply the method of Corollary 21.6. Let y be an element of G which is not a power of the element x of order 2^n. By Proposition 22.3 the automorphism group of $\langle x\rangle$ has four elements of order dividing 2, these being given by

$$x \mapsto x;\ x \mapsto x^{-1};\ x \mapsto x^{2^{n-1}+1};\ \text{and}\ x \mapsto x^{2^{n-1}-1}.$$

The first possibility yields abelian groups. The second automorphism has 1 and $x^{2^{n-1}}$ as fixed points, and so gives the dihedral or generalised quaternion group. The third automorphism fixes each element in $\langle x^2\rangle$. If y^2 were equal to x^{2k}, we would replace y by $x^{-k}y$ if k is even, since this element has order 2. If k is odd, replace y by $x^{-k+2^{n-1}}y$ since

$$\begin{aligned}(x^{-k+2^{n-1}}y)^2 &= x^{-k+2^{n-1}}yx^{-k+2^{n-1}}y \\ &= x^{-k+2^{n-1}}x^{-k+2^{n-1}}y^2 \\ &= x^{-2k}x^{2k}x^{2^n} = 1.\end{aligned}$$

The fourth automorphism fixes the element $z = x^{2^{n-1}}$ of order 2. Thus y^2 could either be 1 (as required) or $x^{2^{n-1}}$. In this latter case, replace y by xy and note that

$$(xy)^2 = xyxy = xyx(y^{-1}z) = xzx^{-1}z = z^2 = 1,$$

as claimed. □

The cases of groups of orders 18, 20, 21, and 28 all require some discussion, but their complete classifications are left as the exercises for this chapter. We shall now outline the process for the next less straightforward case, namely groups with 24 elements.

Proposition 22.10 *A group with 24 elements contains a proper normal subgroup of index at most 3.*

Proof Let G have 24 elements. By Proposition 12.7, G either has a normal subgroup of order 8 (as required) or has a normal subgroup of order 4. In the latter case, the quotient group has order $24/4 = 6$. Since every group of order 6 has a unique Sylow 3-subgroup, we deduce that G has a normal subgroup of index 2 in this case. This completes the proof. □

The above proposition makes it clear how to produce the list of isomorphism classes of groups with 24 elements. Consider the two possibilities that the group has a normal subgroup of index 3 or one of index 2, and use the methods of extension theory on the list of groups of order 8 and groups of order 12, respectively, to complete the classification. There is considerable effort involved, not least in determining the isomorphism types of the constructed groups. The complete list of groups with 24 elements is given in Appendix B.

Similar methods may be used to classify other groups whose orders have a small number of prime divisors. Several examples are suggested in the exercises. However, this method has severe limitations. We present the complete list of isomorphism types of groups of order ≤ 31 in Appendix B. The choice of 31 is not arbitrary: there are 51 isomorphism classes of groups with 32 elements. Even worse, there are 267 isomorphism classes of groups with 64 elements!

Summary for Chapter 22

In this chapter, the techniques of extension theory are used to consider various classification problems. As a preliminary to this, automorphism groups of certain groups are calculated. These include:

- cyclic groups of odd prime order;
- cyclic groups of order 2^n;
- elementary abelian p-groups; and
- the dihedral group $D(3)$.

We then discuss how to combine Sylow theory with the results on extensions to classify groups of certain orders. These include

- groups of order 12;
- groups of order 2^{n+1} with an element of order 2^n;
- groups of order 24; and
- p-groups with only one element of order p (for p an odd prime).

One objective of these discussions is to indicate how to obtain the complete list of non-abelian groups of order less than 32. This list is given in Appendix B.

Exercises 22

1. Show that a non-abelian group with 18 elements is isomorphic to one of $D(9)$, $D(3) \times C_3$ or the group with presentation

$$\langle x, y, z : x^3 = y^3 = z^2 = 1,\ xy = yx,\ zxz^{-1} = x^{-1},\ zyz^{-1} = y^{-1} \rangle.$$

2. Show that a non-abelian group with 208 elements is isomorphic to one of $D(10)$, Q_{20} or $\langle x, y : x^5 = 1 = y^4, \ y^{-1}xy = x^2 \rangle$.

3. Show that a non-abelian group with 28 elements is isomorphic to $D(14)$ or Q_{28}.

4. Find the complete list of groups of order 708 .

5. Which groups G of order 8 are extensions of $C_2 \times C_2$ by C_2?

6. Which of the groups in Proposition 22.8 is isomorphic to $S(3) \times C_2$?

23

The projective special linear groups

In this chapter, we shall consider another class of examples of finite simple groups. In contrast to the alternating groups already discussed, these are not presented as permutation groups, but are produced from matrices over finite fields. We first recall some basic facts about finite fields from Chapter 14.

Proposition 23.1 *Let F be a finite field. There is a prime integer p, the* characteristic *of F, such that F is an elementary abelian p-group under addition. It follows that F has $q = p^k$ elements for some positive integer k. The multiplicative group of F is cyclic of order $q-1$, and so the multiplicative order of any non-zero element of F divides $q-1$. Conversely, for any divisor d of $q-1$, there is an element in F of order d. The number of solutions of the equation $\lambda^n = 1$ in F is the greatest common divisor of n and $q-1$.*

Proof We have already seen in Propositions 14.14 and 14.15 that the additive group of F is an elementary abelian p-group, and that the multiplicative group is cyclic of order $q-1$. We also know from Proposition 4.14 that a cyclic group of order n has elements of order d for every divisor d of n. It only remains to find the number of solutions of the equation $\lambda^n = 1$ in F.

Suppose first that t is an integer with no divisors in common with $q-1$ (apart from 1). It follows from a standard fact in number theory (see Theorem A.2 of Appendix A) that there are integers r and s such that

$$rt + s(q-1) = 1.$$

Thus, if $\mu \in F$ satisfies the equation $\mu^t = 1$, then since $\mu^{q-1} = 1$,

$$\mu = \mu^1 = \mu^{tr}\mu^{(q-1)s} = 1.$$

Hence if t has no divisors in common with $q-1$, the only solution of $\lambda^t = 1$ in F is $\lambda = 1$.

Now, for any integer n, we write $n = dm$, where d is the greatest common divisor of n and $q-1$, and $m = n/d$. It follows that m and $q-1$ have no common divisors other than 1. The solutions of $\lambda^n = 1$ in F are the solutions of $(\lambda^d)^m = 1$. The previous discussion shows that these solutions are precisely the solutions in F of $\lambda^d = 1$. By Proposition 4.14, if d is a divisor of the order of a cyclic group, the number of solutions of the equation $x^d = 1$ in F is d, and so the result follows. □

We next recall a definition.

Definition 23.2 *For any positive integer k and any prime integer p, let F be a finite field of order $q = p^k$. For any positive integer n, the* general linear group, *$GL(n,q)$, is the set of all invertible $n \times n$ matrices over F under matrix multiplication.*

Proposition 23.3 *The order of the group $GL(n,q)$ is*

$$(q^n - 1)(q^n - q)(q^n - q^2)\dots(q^n - q^{n-1}).$$

Proof We need to count the number of $n \times n$ matrices with non-zero determinant. We do this by considering each possible column of the matrix A as discussed in Appendix A. If A is to have non-zero determinant, the first column of A cannot be the zero vector. There are therefore $q^n - 1$ possible choices for the first column. If the second column were a multiple of the first, subtracting this multiple of the first from the second column would give a matrix with a column of zeroes having the same determinant as A. There are therefore $q^n - q$ choices for the second column. Continuing in this way, it can be seen that the ith column of A can be chosen in $q^n - q^{i-1}$ ways in order that this column is not a linear combination of the previous $i-1$ columns. This proves the result. □

Example 23.4 We saw in Example 22.5 that the group $GL(2,2)$ has 6 elements and is in fact isomorphic to the symmetric group $S(3)$. The group $GL(2,3)$ has 48 elements. The group $GL(2,4)$ has 180 elements.

Definition 23.5 *The* special linear group *$SL(n,q)$ is the subgroup of the group $GL(n,q)$ consisting of those matrices of determinant 1.*

Remark We saw in Proposition 14.14 that any finite field has order q for some prime power $q = p^k$. In fact, any two finite fields with the same number of elements are isomorphic. This field is usually denoted $GF(q)$ with its multiplicative group of non-zero elements being denoted

$GF(q)^\times$. We have also seen that $SL(n,q)$ is the kernel of the homomorphism $GL(n,q) \to GF(q)^\times$ defined by $A \mapsto \det(A)$. Since there are $q-1$ possible non-zero determinants, the Homomorphism Theorem shows that the index of $SL(n,q)$ in $GL(n,q)$ is $q-1$ and that $SL(n,q)$ is a normal subgroup with the quotient group $GL(n,q)/SL(n,q)$ isomorphic to the multiplicative group $GF(q)^\times$.

Example 23.6 There are 2 non-zero elements in the field with 3 elements, so $SL(2,3)$ has index 2 in $GL(2,3)$, and so has order 24. There are three possible non-zero determinants in the field with 4 elements, and so $SL(2,4)$ has 60 elements.

Definition 23.7 *For any positive integer n, the $n \times n$* scalar matrices *over $GF(q)$ are those matrices in $SL(n,q)$ which are of the form λI_n for some $\lambda \in GF(q)^\times$.*

Remark The scalar matrices commute with each matrix and so any subgroup consisting of scalar matrices is a normal subgroup of any subgroup of $GF(n,q)$ containing the subgroup.

Definition 23.8 *The* projective special linear group $PSL(n,q)$ *is the quotient group $SL(n,q)/Z$, where Z is the subgroup of scalar matrices in $SL(n,q)$.*

Remark If λI_n has determinant 1, then $\lambda^n = 1$, and so the number of elements in Z is, by Proposition 23.1, the greatest common divisor d, say, of n and $q-1$. Thus $PSL(n,q)$ has order

$$(q^n-1)(q^n-q)(q^n-q^2)\dots(q^n-q^{n-1})/(q-1)d.$$

In particular, when $n=2$ and p is an odd prime,

$$|PSL(2,q)| = q(q^2-1)/2.$$

Example 23.9 The group $PSL(2,3)$ has order $24/2 = 12$. In fact, the group $PSL(2,3)$ is isomorphic to the alternating group $A(4)$. To see this consider the following two matrices of determinant 1 over $\mathbf{Z}_3$:

$$A = \begin{pmatrix} 0 & 2 \\ 1 & 0 \end{pmatrix}; \; B = \begin{pmatrix} 1 & 1 \\ 1 & 2 \end{pmatrix}.$$

Since $A^2 = B^2 = -I$ and $BAB^{-1} = A^{-1}$, these matrices satisfy the relations for the quaternion group of order 8. Now let X be the matrix $\begin{pmatrix} 1 & 1 \\ 0 & 1 \end{pmatrix}$, so that $X^3 = I$. It may be checked that

$$XAX^{-1} = B \quad \text{and} \quad XBX^{-1} = AB.$$

Since $\langle A, B, X\rangle$ is a subgroup of $SL(2,3)$ which is a cyclic extension of a group of order eight by an element of order three, $\langle A, B, X\rangle$ must equal $SL(2,3)$.

Now let Z denote the set of scalar matrices in $SL(2,3)$, so that $Z = \{I_2, -I_2\}$. Then the quotient group $SL(2,3)/Z$ is isomorphic to $A(4)$ by the map specified by

$$\vartheta(AZ) = (12)(34); \quad \vartheta(BZ) = (14)(23); \quad \text{and} \quad \vartheta(XZ) = (123).$$

Example 23.10 In the field with four elements, each non-zero element satisfies the equation $\lambda^3 = 1$, and so the only 2×2 scalar matrix of determinant 1 is I. It follows that $PSL(2,4) = SL(2,4)$. This fact generalises to any field with an even number of elements.

We next show that $PSL(2,5)$ is simple using a method similar to that used in the proof that $A(5)$ is simple in Theorem 16.13.

Proposition 23.11 *The group $PSL(2,5)$ is a non-abelian simple group.*

Proof It is easy to see that $PSL(2,5)$ is non-abelian since

$$\begin{pmatrix} 1 & 1 \\ 0 & 1 \end{pmatrix} Z \begin{pmatrix} 0 & 1 \\ 4 & 4 \end{pmatrix} Z = \begin{pmatrix} 4 & 4 \\ 0 & 4 \end{pmatrix} Z \neq \begin{pmatrix} 0 & 1 \\ 4 & 4 \end{pmatrix} Z \begin{pmatrix} 1 & 1 \\ 0 & 1 \end{pmatrix} Z.$$

We shall show that the only normal subgroups of $G = SL(2,5)$ are: G itself; $\{I, -I\}$ and $\{I\}$. It will then follow by the Correspondence Theorem that the quotient group $PSL(2,5)$ is a non-abelian simple group.

The first step is to determine the conjugacy classes of G. The elements I and $-I$ are both central, and so each forms a conjugacy class on its own.

We now calculate the conjugacy class of the element $T = \begin{pmatrix} 0 & 4 \\ 1 & 0 \end{pmatrix}$ of order 4, by determining the centraliser of T. This consists of those matrices

$$\begin{pmatrix} a & b \\ c & d \end{pmatrix}$$

in G such that

$$\begin{pmatrix} 0 & 4 \\ 1 & 0 \end{pmatrix}\begin{pmatrix} a & b \\ c & d \end{pmatrix} = \begin{pmatrix} a & b \\ c & d \end{pmatrix}\begin{pmatrix} 0 & 4 \\ 1 & 0 \end{pmatrix}.$$

This yields the equations

$$4c = b; \quad a = d; \quad 4d = 4a \quad \text{and} \quad b = 4c.$$

Thus the centraliser of T consists of matrices of the form $\begin{pmatrix} a & 4c \\ c & a \end{pmatrix}$ with $a^2 + c^2 = 1$. Since 0, 1 and 4 are the only squares in $\mathbf{Z}_5$, it may easily be seen that there are four elements in the centraliser of T (these being I, T, T^2 and T^3). Thus T has $120/4 = 30$ conjugates.

Similarly, we can show that the centraliser in G of $S = \begin{pmatrix} 0 & 1 \\ 4 & 4 \end{pmatrix}$ of order 3 consists of the six matrices $\{I, S, S^2, -I, -S, -S^2\}$. It follows that S has 20 conjugates in G. It may be seen that the centraliser of $-S$ is identical to the centraliser of S. Since $-S$ is not in the conjugacy class of S (the matrices S and $-S$ have different orders), $-S$ will also have 20 conjugates.

We next consider the conjugacy class of $R = \begin{pmatrix} 1 & 1 \\ 0 & 1 \end{pmatrix}$ of order 5. An element in the centraliser of R is of the form $\begin{pmatrix} a & b \\ c & d \end{pmatrix}$, where

$$\begin{pmatrix} 1 & 1 \\ 0 & 1 \end{pmatrix}\begin{pmatrix} a & b \\ c & d \end{pmatrix} = \begin{pmatrix} a & b \\ c & d \end{pmatrix}\begin{pmatrix} 1 & 1 \\ 0 & 1 \end{pmatrix}.$$

This gives the equations

$$a + c = a; \quad c = c; \quad b + d = a + b \quad \text{and} \quad d = c + d.$$

Thus the matrices which commute with R are of the form $\begin{pmatrix} a & b \\ 0 & a \end{pmatrix}$. If this matrix has determinant 1, we see that a is 1 or 4. Since b can then be any element of $\mathbf{Z}_5$, the centraliser of R has 10 elements, and so R has 12 conjugates. It may be checked that R^2 is not conjugate to R, but that R^2 has the same centraliser as that of R. Thus R^2 also has 12 conjugates. Similarly, the matrix $-R$, of order 10, has the same centraliser, as does $-R^2$, so these also have 12 conjugates each.

We have therefore found the following conjugacy classes in G:

$\mathcal{C}_1 = \{I\}$;
$\mathcal{C}_2 = \{-I\}$;
$\mathcal{C}_3 = \{30 \text{ conjugates of an element } T = \begin{pmatrix} 0 & 4 \\ 1 & 0 \end{pmatrix} \text{ of order } 4\}$;
$\mathcal{C}_4 = \{20 \text{ conjugates of } S = \begin{pmatrix} 0 & 1 \\ 4 & 4 \end{pmatrix} \text{ of order } 3\}$;

$C_5 = \{20 \text{ conjugates of } -S \text{ of order } 6\};$
$C_6 = \{12 \text{ conjugates of } R = \begin{pmatrix} 1 & 1 \\ 0 & 1 \end{pmatrix} \text{ each of order } 5\};$
$C_7 = \{12 \text{ squares of the elements in } C_6\};$
$C_8 = \{12 \text{ elements of order 10 of the form } -R \text{ for } R \in C_6\};$ and
$C_9 = \{12 \text{ elements of order 10, of the form } -R^2 \text{ for } R \in C_6\}.$

Since these classes contain 120 matrices, they must be the complete list of conjugacy classes of G.

The proof of simplicity then proceeds as follows. Let N be a normal subgroup of G which contains $\{I, -I\}$, and suppose that $N \neq \{I, -I\}$. Suppose that N contains an element of order 5. Since N is a union of conjugacy classes and contains $-I$ and N is also closed under multiplication, this subgroup must contain each of C_6, C_7, C_8 and C_9. Thus N contains at least $2 + 4(12) = 50$ elements. Since there is no conjugacy class with 10 elements, and the number of elements in a subgroup of G divides 120, the only possibility for N in this case is that N is equal to G itself.

We may therefore suppose that N has no elements of order 5 and so $|N|$ divides 24 . Since N contains I and $-I$ and is a union of conjugacy classes, we see that the only possibility is for N to equal $\{I, -I\}$. We have shown that the only normal subgroups of G containing $Z = \{I, -I\}$ are Z and G, so that the quotient group G/Z is simple. □

Remark A similar proof may also be used to show that $PSL(2, 4)$ is simple, although this is easier than the proof of Proposition 23.11 since $PSL(2, 4) = SL(2, 4)$. Let ω be an element of the field with four elements other than 0 or 1, so that $\omega^3 = 1$.

Thus if G denotes the group $SL(2, 4)$ of order 60, the conjugacy classes of G are

$C_1 = \{I\};$
$C_2 = \{15 \text{ elements of order 2, all conjugate to } \begin{pmatrix} 0 & 1 \\ 1 & 0 \end{pmatrix}\};$
$C_3 = \{20 \text{ elements conjugate to } \begin{pmatrix} \omega & 0 \\ 0 & \omega^2 \end{pmatrix}, \text{ all of order } 3\};$
$C_4 = \{12 \text{ elements of order 5, all conjugate to } R = \begin{pmatrix} 1 & \omega \\ \omega & \omega \end{pmatrix}\};$ and
$C_5 = \{12 \text{ elements of order 5, all conjugate to } R^2\}.$

Since these classes contain a total of 60 elements, this must be the complete list of conjugacy classes of G. The proof that G is simple now follows exactly as in the proof for the simplicity of $A(5)$. There is no subset of the set of conjugacy classes of G which (a) contains C_1 and (b) has cardinality dividing 60, except for the subsets $\{I\}$ and G.

Remark We have now seen three simple groups with 60 elements, these being $A(5), PSL(2,4)$ and $PSL(2,5)$. It may be shown that these three groups are isomorphic.

Proposition 23.12 *Each of the groups $PSL(2,4)$ and $PSL(2,5)$ is isomorphic to the alternating group $A(5)$.*

Proof Each of the groups $PSL(2,4)$ and $PSL(2,5)$ has a unique conjugacy class of 15 elements of order 2, and no elements of order 4. It can be seen that these 15 elements form (together with the identity element) five subgroups of order 4. Since 4 is the order of a Sylow 2-subgroup, we deduce that the normaliser of any Sylow 2-subgroup has index 5. Now apply Proposition 9.22 to the simple group in question (so that G is either $PSL(2,4)$ or $PSL(2,5)$) with H as the normaliser of a Sylow 2-subgroup. Since G has no non-trivial normal subgroups, we conclude that G is isomorphic to a subgroup of $S(5)$. Since any subgroup of $S(5)$ of order 60 would have index 2, and so be normal, the Jordan–Hölder Theorem now implies that G is isomorphic to $A(5)$, as required. □

Remark The conclusion of Proposition 23.12 holds for any simple group with 60 elements. The method of proof shows that we only need to prove that such a group has 15 Sylow 2-subgroups.

The main result of this section is the fact that, except when $q < 4$ the group $PSL(2,q)$ is simple for all prime powers $q = p^k$. We have seen that the groups $PSL(2,2)$ and $PSL(2,3)$ are genuine exceptions to this result in that both are soluble groups, and also that $PSL(2,4)$ and $PSL(2,5)$ are indeed simple. Several preliminaries are needed before we give the main proof.

Definition 23.13 *An element of $GL(n,q)$ is a* transvection *if it is of the form $B_{i,j}(\lambda) = I + E_{i,j}(\lambda)$, where I is the identity $n \times n$ matrix and $E_{i,j}(\lambda)$ is an elementary matrix (a matrix with one non-zero entry equal to λ in location (i,j)). Any transvection has determinant 1 and so is in $SL(n,q)$.*

The importance of these special types of matrices is that, for any matrix A, the product $B_{i,j}(\lambda)A$ is the matrix obtained from A by adding λ times the jth row of A to the ith row. Thus, for example,

$$B_{2,3}(2) = \begin{pmatrix} 1 & 2 & 3 \\ 1 & 0 & 1 \\ 2 & 1 & 0 \end{pmatrix} = \begin{pmatrix} 1 & 0 & 0 \\ 0 & 1 & 2 \\ 0 & 0 & 1 \end{pmatrix} \begin{pmatrix} 1 & 2 & 3 \\ 1 & 0 & 1 \\ 2 & 1 & 0 \end{pmatrix} = \begin{pmatrix} 1 & 2 & 3 \\ 5 & 2 & 1 \\ 2 & 1 & 0 \end{pmatrix}.$$

This means that we can duplicate any row reduction of any matrix A by multiplying A on the left by a sequence of transvections.

Proposition 23.14 *Every element A of $GL(2,q)$ can be written as a product TD where T is a product of transvections and D is the matrix*

$$\begin{pmatrix} 1 & 0 \\ 0 & \det(A) \end{pmatrix}.$$

In particular, each element of $SL(2,q)$ is a product of transvections.

Proof Suppose first that the (1,1) entry of A is zero. Since A is invertible, the (2,1) entry cannot also be zero. In this case, therefore, adding the second row to the first will produce a matrix B whose (1,1) entry is non-zero. Now define a matrix A_1 to equal A if the (1, 1) entry of A is non-zero, and define A_1 to be $B = B_{1,2}(1)A$ if A has zero entry in the (1,1) place. It follows that in either case $\det(A_1) = \det(A)$. Now A has the form

$$A_1 = \begin{pmatrix} a & b \\ c & d \end{pmatrix}, \text{ with } a \text{ non-zero.}$$

Then adding a suitable multiple of the first row of the matrix A_1 to its second row produces a matrix whose (2,1) entry is zero. Then $B_{2,1}(-c/a)A_1$ is a matrix of the form

$$A_2 = \begin{pmatrix} \alpha & \beta \\ 0 & \delta \end{pmatrix}, \text{ with } \alpha \text{ non-zero.}$$

Since A_2 has determinant equal to $\det(B_{2,1}(-c/a)A_1) = \det(A_1)$, it follows that δ is also non-zero. Now, adding a suitable multiple of the second row to the first produces a diagonal matrix

$$D = \begin{pmatrix} \lambda & 0 \\ 0 & \mu \end{pmatrix} = B_{1,2}(-\beta/\delta)A_2,$$

where $\lambda\mu = \det(A)$. Finally, note that we can reduce D by a series of row operations (multiplication on the left by transvections) as follows:

(1) add the first row to the second to obtain $\begin{pmatrix} \lambda & 0 \\ \lambda & \mu \end{pmatrix}$;

(2) add $(1-\lambda)/\lambda$ times the second row to the first to obtain $\begin{pmatrix} 1 & \nu \\ \lambda & \mu \end{pmatrix}$,

for some ν;

(3) subtract λ times the first row from the second to obtain $\begin{pmatrix} 1 & \rho \\ 0 & \sigma \end{pmatrix}$,

where $\sigma = \det(A)$; finally

(4) subtract ρ/σ times the second row from the first to obtain $\begin{pmatrix} 1 & 0 \\ 0 & \sigma \end{pmatrix}$,

as required.

When A is in $SL(2,q)$, we know $\sigma = \det(A) = 1$, and so A is a product of transvections. □

We now come to the main result of this section. The proof we give of this is essentially that given by Dickson (1900). This result uses the following elementary fact from the general theory of vector spaces, discussed in Appendix A. If v and w are column vectors of length 2 over $GF(q)$, and neither is a scalar multiple of the other, then the vectors are a basis for the vector space of all such column vectors. Thus, any column vector of length 2 over $GF(q)$ can be written in the form $av + bw$ for some $a, b \in GF(q)$.

Theorem 23.15 *The group $PSL(2, q)$ is simple if $q > 3$.*

Proof As in the case when $q = 5$ (Proposition 23.11), it is sufficient to show that the only normal subgroup of $SL(2, q)$ which contains a matrix which is not a scalar matrix is $SL(2, q)$ itself. Therefore, let N be a normal subgroup of $SL(2, q)$ which contains an element A which is not a scalar matrix. The first objective is to show that it is then impossible for Av always to be a multiple of v. To see this, let

$$A = \begin{pmatrix} a & b \\ c & d \end{pmatrix} \text{ and } e_1 = \begin{pmatrix} 1 \\ 0 \end{pmatrix}, \ e_2 = \begin{pmatrix} 0 \\ 1 \end{pmatrix} \text{ and } e_3 = \begin{pmatrix} 1 \\ 1 \end{pmatrix}.$$

Then if $Ae_1 = \lambda e_1$, we see that $c = 0$, if $Ae_2 = \mu e_2$ then $b = 0$, and finally if $Ae_3 = \nu e_3$, it follows that $a = d$, so that A is a scalar matrix, contrary to assumption. This means that there is a column vector v such that $Av = w$ is not a multiple of v. By the elementary fact quoted before this proof, writing T for the matrix whose columns are v and w, we have shown that T has non-zero determinant. This means that we can find solutions a, b to the matrix equation

$$T\begin{pmatrix} a \\ b \end{pmatrix} = \begin{pmatrix} r \\ s \end{pmatrix}, \text{ where } Aw = \begin{pmatrix} r \\ s \end{pmatrix},$$

namely

$$\begin{pmatrix} a \\ b \end{pmatrix} = T^{-1}Aw.$$

Since

$$T\begin{pmatrix} 0 \\ 1 \end{pmatrix} = w = Av \text{ and } T\begin{pmatrix} a \\ b \end{pmatrix} = Aw,$$

we have shown that

$$AT = TB, \text{ where } B = \begin{pmatrix} 0 & a \\ 1 & b \end{pmatrix}.$$

It follows that $B = T^{-1}AT$. Furthermore, since A has determinant 1 so does B, and hence $a = -1$. If $\det(T) \neq 1$, we may replace N by $T^{-1}NT$

and $SL(2,q)$ by $T^{-1}SL(2,q)T = SL(2,q)$ to obtain a normal subgroup of $SL(2,q)$ which contains the matrix B, where

$$B = T^{-1}AT = \begin{pmatrix} 0 & -1 \\ 1 & b \end{pmatrix}.$$

Then we obtain another element of N from the product $(C^{-1}B^{-1}C)B$, where C is a diagonal matrix with entries c and c^{-1} and c is any non-zero element of $GF(q)$. In fact

$$\begin{pmatrix} c & 0 \\ 0 & c^{-1} \end{pmatrix}^{-1} \begin{pmatrix} 0 & -1 \\ 1 & b \end{pmatrix}^{-1} \begin{pmatrix} c & 0 \\ 0 & c^{-1} \end{pmatrix} \begin{pmatrix} 0 & -1 \\ 1 & b \end{pmatrix} = \begin{pmatrix} c^{-2} & \gamma \\ 0 & c^2 \end{pmatrix},$$

where $\gamma = b(c^{-2} - 1)$. We obtain a further element of N for each choice of μ from the product

$$\begin{pmatrix} 1 & \mu \\ 0 & 1 \end{pmatrix}^{-1} \begin{pmatrix} c^{-2} & \gamma \\ 0 & c^2 \end{pmatrix}^{-1} \begin{pmatrix} 1 & \mu \\ 0 & 1 \end{pmatrix} \begin{pmatrix} c^{-2} & \gamma \\ 0 & c^2 \end{pmatrix} = \begin{pmatrix} 1 & \mu(1-c^4) \\ 0 & 1 \end{pmatrix}.$$

It follows that if there is an element c of $GF(q)$ with $c^4 \neq 1$, then we can ensure that every transvection of the form $\begin{pmatrix} 1 & \lambda \\ 0 & 1 \end{pmatrix}$ is in the subgroup N, by suitable choice of c and μ. Once we know this, we observe that

$$\begin{pmatrix} 0 & -1 \\ 1 & 0 \end{pmatrix}^{-1} \begin{pmatrix} 1 & \lambda \\ 0 & 1 \end{pmatrix} \begin{pmatrix} 0 & -1 \\ 1 & 0 \end{pmatrix} = \begin{pmatrix} 1 & 0 \\ -\lambda & 1 \end{pmatrix},$$

and so every transvection is in N, so that $N = SL(2,q)$ by Proposition 23.14.

By Proposition 23.1, the only q greater than 3 such that every element in the field with q elements satisfies $c^4 = 1$ is $q = 5$. In this case the group is simple by Proposition 23.11. This completes the proof. □

Proposition 23.16 *Let q be an odd prime power. The matrices I and $-I$ are the only central elements in $SL(2,q)$.*

Proof Suppose that

$$W = \begin{pmatrix} a & b \\ c & d \end{pmatrix}$$

is in the centre of $SL(2,q)$. The fact that W commutes with the transvection $B_{1,2}(1)$ shows that $c = 0$ and $a = d$. The fact that W also commutes with $B_{2,1}(1)$ then shows that $b = 0$, so that W is the scalar matrix, aI. Since W has determinant $1, a^2 = 1$, and so a is either 1 or -1. □

Remark The results of this chapter can be generalised to matrices of arbitrary size. In fact for any $n > 2$ and for any prime power q, the group $PSL(n,q) = SL(n,q)/Z$, where Z is the set of scalar matrices of determinant 1, is a non-abelian simple group. The text by Suzuki (1982) gives full details of this.

Summary for Chapter 23

This chapter is devoted to the proof of one main result which gives further examples of finite simple groups. Let $F = GF(q)$ be a finite field with more than 3 elements. Then the group $SL(2,q)$ of 2×2 matrices of determinant 1 over $GF(q)$ has a non-abelian simple quotient group $PSL(2,q) = SL(2,q)/Z$, where Z is $\{I\}$ if q is even and Z is $\{I, -I\}$ if q is odd.

The proof of this result and its generalisation to larger matrices was given in Jordan's *Traité des Substitutions* (1870), for the case when the field F is a prime field $\mathbf{Z}_p$. The generalisation to arbitrary finite fields was made by Dickson.

24

The Mathieu groups

In this chapter, we shall construct the Mathieu group, and briefly discuss the other Mathieu groups. The group M_{11} is another example of a finite simple group. However, this group is not one of an infinite family of examples, but is one of the *sporadic* simple groups. The sporadic groups comprise 26 groups which are not members of any of the infinite families of finite simple groups.

We shall define M_{11} in three steps, the first of these being to define a group M_9 of order 72.

The automorphism group of $N = C_3 \times C_3$ contains the subgroup Q generated by the following matrices over $\mathbf{Z}_3$:

$$A = \begin{pmatrix} 0 & 2 \\ 1 & 0 \end{pmatrix} \text{ and } B = \begin{pmatrix} 1 & 1 \\ 1 & 2 \end{pmatrix}.$$

Since $A^2 = B^2 = -I$ and $BAB^{-1} = A^{-1}$, we see that Q is isomorphic to the quaternion group of order 8. We may therefore form the semidirect product of N by Q. This subgroup of the holomorph of N is a group of order 72 which we denote by M_9.

It is also possible to represent the elements of this group as permutations in $S(9)$ as follows: let π_1 and π_2 be the permutations

$$\pi_1 = (1\ 2\ 3)(4\ 5\ 6)(7\ 8\ 9),$$

$$\pi_2 = (1\ 4\ 7)(2\ 5\ 8)(3\ 6\ 9).$$

Since $\pi_1^3 = 1 = \pi_2^3$ and $\pi_1\pi_2 = \pi_2\pi_1$, the group generated by π_1 and π_2 is isomorphic to N. Now let

$$\rho_1 = (2\ 4\ 3\ 7)(5\ 6\ 9\ 8), \quad \text{and} \quad \rho_2 = (2\ 5\ 3\ 9)(4\ 8\ 7\ 6).$$

It may be checked that

$$\rho_1^2 = (2\ 3)(4\ 7)(5\ 9)(6\ 8) = \rho_2^2 \quad \text{and} \quad \rho_2\rho_1\rho_2^{-1} = \rho_1^{-1}.$$

Hence $\langle\rho_1, \rho_2\rangle$ is isomorphic to the quaternion group of order 8. Then

$$\rho_1\pi_1\rho_1^{-1} = \pi_2; \quad \rho_1\pi_2\rho_1^{-1} = \pi_1^2;$$

$$\rho_2\pi_1\rho_2^{-1} = \pi_1\pi_2; \quad \text{and } \rho_2\pi_2\rho_2^{-1} = \pi_1\pi_2^2,$$

so that ρ_1 and ρ_2 act on $\langle\pi_1, \pi_2\rangle$ in precisely the way in which the matrices A and B act on $C_3 \times C_3$. This shows that $\langle\pi_1, \pi_2, \rho_1, \rho_2\rangle$ is isomorphic to NQ, as required.

Definition 24.1 *A subgroup G of a symmetric group on a set X is* transitive *if for any element $x \in X$, the orbit of x is X.*

Remark Notice that when G is transitive on X, the stabiliser G_x has index $|X|$ in G, and that each $y \in X$ is of the form $g \cdot x$, for some $g \in G$. It follows easily from this that the orbit of any element y of X is the whole of X.

Proposition 24.2 *With the above notation, $G = \langle\pi_1, \pi_2, \rho_1, \rho_2\rangle$ is a transitive group of order 72 isomorphic to M_9. The stabiliser of 1 is the subgroup $Q = \langle\rho_1, \rho_2\rangle$ of order 8. If τ is any element of G not in Q, then $G = Q \cup Q\tau Q$, where $Q\tau Q$ denotes the double coset $\{x\tau y : x, y \in Q\}$.*

Proof It follows directly from the definition that G is isomorphic to M_9, so that G has 72 elements. It follows from its definition that the subgroup $\langle\pi_1, \pi_2\rangle$ is transitive on the set $\{1, \ldots, 9\}$, and hence that G is also transitive on this set. It is clear that Q is transitive on the set $\{2, \ldots, 9\}$. Since G has 72 elements, the Orbit–Stabiliser Theorem shows that G_1, the stabiliser of 1, has index 9 in G. Since $|G_1| = 8$ and Q is a subgroup of G, it follows that $G_1 = Q$. For any element τ of G not in Q, the integer $\tau(1)$ is equal to some i with $i \neq 1$, and so $\tau^{-1}(1)$ is in the set $\{2, \ldots, 9\}$. Then, if ϑ is another permutation not in Q, the integer $\vartheta^{-1}(1)$ is also in the set $\{2, \ldots, 9\}$. Since Q is transitive on this set, there is a permutation ϕ in Q taking $\tau^{-1}(1)$ to $\vartheta^{-1}(1)$. Thus, $\phi\tau^{-1}(1) = \vartheta^{-1}(1)$, so that $\vartheta\varphi\tau^{-1}(1) = 1$. Since Q is the stabiliser of 1, the permutation $\vartheta\varphi\tau^{-1} \in Q$, so that $\vartheta\varphi \in Q\tau$ and $\vartheta \in Q\tau Q$, as required. □

The next step is to define the group M_{10} of order 720.

Proposition 24.3 *Let M_{10} be the subgroup of $S(10)$ generated by M_9 together with the permutation*

$$\sigma = (1\ 10)(4\ 5)(6\ 8)(7\ 9).$$

Then $M_{10} = M_9 \cup M_9\sigma M_9$ is a transitive group of order 720 in which the stabiliser of 10 is the group M_9.

Proof Denote the subgroup M_9 by M. We show that $S = M \cup M\sigma M$ is closed under multiplication. To do this, let $x, y \in S$. There are four cases to consider:

(1) $x, y \in M$;
(2) $x \in M$ and $y \in M\sigma M$;
(3) $x \in M\sigma M$ and $y \in M$; and
(4) $x, y \in M\sigma M$.

In the first case $xy \in M$ since M is a group; in the second and third cases $xy \in M\sigma M$. In the last case, $xy \in M\sigma M\sigma M$, so we must show that $\sigma M\sigma$ is a subset of S. To do this, we use Proposition 24.2 and write M as $Q \cup Q\tau Q$ for the specific element

$$\tau = \pi_1\rho_1^2\pi_1^{-1} = (1\ 3)(4\ 9)(5\ 8)(6\ 7)$$

of $M\backslash Q$. Since $\sigma^2 = 1$,

$$\begin{aligned}\sigma\rho_1\sigma &= \sigma(2\ 4\ 3\ 7)(5\ 6\ 9\ 8)\sigma^{-1} \\ &= (2\ 5\ 3\ 9)(4\ 8\ 7\ 6) = \rho_2,\end{aligned}$$

and

$$\begin{aligned}\sigma\rho_2\sigma &= \sigma(2\ 5\ 3\ 9)(4\ 8\ 7\ 6)\sigma^{-1} \\ &= (2\ 4\ 3\ 7)(5\ 6\ 9\ 8) = \rho_1.\end{aligned}$$

It follows that $\sigma Q\sigma = Q$, and so $\sigma Q = Q\sigma$. It only remains to show that $\sigma(Q\tau Q)\sigma \subseteq S$. This follows because

$$\sigma\tau = (1\ 3\ 10)(4\ 7\ 8)(5\ 6\ 9),$$

so that $(\sigma\tau)^3 = 1$ and $\sigma\tau\sigma = \tau\sigma\tau$. Hence

$$\sigma(Q\tau Q)\sigma = Q(\sigma\tau\sigma)Q, = Q\tau\sigma\tau Q \subseteq M\sigma M \subseteq S,$$

as required.

We have therefore shown that $S = M \cup M\sigma M$ is a group. Since S contains M and σ, it follows that $\langle M, \sigma\rangle \subseteq S$. However, $\langle M, \sigma\rangle$ must clearly contain M and $M\sigma M$, so $S \subseteq \langle M, \sigma\rangle$. Thus,

$$M_{10} = \langle M, \sigma\rangle = M \cup M\sigma M,$$

as required. It is clear that M_{10} is transitive since, in the first place, it contains M, and so is transitive on $\{1, \ldots, 9\}$. Also M_{10} contains all products of the form $\mu\sigma$, as μ ranges over M_9, so that 10 can be sent to any integer in $\{1, \ldots, 10\}$ by a suitable element of M_{10}. Every element in M fixes 10, but no element in $M\sigma M$ fixes 10, since for μ_1, μ_2 in M,

$$\mu_1\sigma\mu_2(10) = \mu_1\sigma(10) = \mu_1(1) \in \{1, \ldots, 9\}.$$

Hence M is the stabiliser of 10 in $M_{10} = M_9 \cup M_9\sigma M_9$. It follows by the Orbit–Stabiliser Theorem that $|M_{10}| = 10|M_9| = 720$, as required. □

This argument may be repeated to show the following.

Proposition 24.4 *Let M_{11} be the subgroup of $S(11)$ generated by M_{10} and the permutation*

$$\nu = (4\ 7)(5\ 8)(6\ 9)(10\ 11).$$

Then $M_{11} = M_{10} \cup M_{10}\nu M_{10}$ is a transitive group of order 7920 in which the stabiliser of 11 is the group M_{10}.

Proof Denoting the subgroup M_{10} by S, we show that $T = S \cup S\nu S$ is closed under multiplication. As in the proof of Proposition 24.3, there are four cases to consider, but three of these are straightforward. The remaining case arises when $x, y \in S\nu S$ and we need to show that $xy \in T$. We do this by showing that $\nu S\nu$ is a subset of T.

We now use Proposition 24.3, recalling that $S = M \cup M\sigma M$, where $\sigma = (1\ 10)(4\ 5)(6\ 8)(7\ 9)$. Since $\nu^2 = 1$,

$$\begin{aligned} \nu\rho_1\nu &= \nu(2\ 4\ 3\ 7)(5\ 6\ 9\ 8)\nu^{-1} \\ &= (2\ 7\ 3\ 4)(5\ 8\ 9\ 6) = \rho_1^{-1} \end{aligned}$$

and

$$\begin{aligned} \nu\rho_2\nu &= \nu(2\ 5\ 3\ 9)(4\ 8\ 7\ 6)\nu^{-1} \\ &= (2\ 8\ 3\ 6)(4\ 9\ 7\ 5) = \rho_1^{-1}\rho_2. \end{aligned}$$

Also

$$\begin{aligned} \nu\pi_1\nu &= \nu(1\ 2\ 3)(4\ 5\ 6)(7\ 8\ 9)\nu^{-1} \\ &= (1\ 2\ 3)(7\ 8\ 9)(4\ 5\ 6) = \pi_1 \end{aligned}$$

and

$$\begin{aligned} \nu\pi_2\nu &= \nu(1\ 4\ 7)(2\ 5\ 8)(3\ 6\ 9)\nu^{-1} \\ &= (1\ 7\ 4)(2\ 8\ 5)(3\ 9\ 6) = \pi_2^{-1}. \end{aligned}$$

Thus $\nu M\nu = M \subseteq S \subseteq T$, which completes the first part of the proof that $\nu S\nu \subseteq T$.

The next step is to show that $\nu(M\sigma M)\nu$ is a subset of T. This follows because

$$\begin{aligned} \nu\sigma &= (4\ 7)(5\ 8)(6\ 9)(10\ 11)(1\ 10)(4\ 5)(6\ 8)(7\ 9) \\ &= (1\ 11\ 10)(4\ 8\ 9)(5\ 7\ 6), \end{aligned}$$

so that $(\nu\sigma)^3 = 1$ and $\nu\sigma\nu = \sigma\nu\sigma$. Since $\nu M\nu = M$, it follows that $\nu M = M\nu$, and so

$$\nu(M\sigma M)\nu = M(\nu\sigma\nu)M, = M\sigma\nu\sigma M \subseteq S\nu S \subseteq T,$$

as required.

As in the proof of Proposition 24.3, it follows that

$$M_{11} = T = \langle S, \nu\rangle = S \cup S\nu S.$$

It is clear that M_{11} is transitive since it contains all elements of $S = M_{10}$, and so is transitive on $\{1, \ldots, 10\}$ and also contains all products of the form $\mu\nu$, so that 11 can be sent to any integer in $\{1, \ldots, 11\}$ by a suitable choice of the element μ of M_{10}. Every element in M_{10} fixes 11, but no element in $M_{10}\nu M_{10}$ fixes 11, so M_{10} is the stabiliser of 11 in the group $M_{11} = M_{10} \cup M_{10}\nu M_{10}$. It again follows by the Orbit–Stabiliser Theorem that $|M_{11} : M_{10}| = 11$ so that $|M_{11}| = 7920$, as required. □

Remark 1 It will be shown in Theorem 24.6 below that M_{11} is a non-abelian simple group.

Remark 2 The method of proof given above (sometimes known as transitive extension) may be applied once more, by defining the permutation $\vartheta = (4\ 9)(5\ 7)(6\ 8)(11\ 12)$, and setting M_{12} to be the group generated by M_{11} and ϑ. Since $\vartheta\nu$ is the permutation

$$(4\ 9)(5\ 7)(6\ 8)(11\ 12)(4\ 7)(5\ 8)(6\ 9)(10\ 11) = (4\ 5\ 6)(7\ 9\ 8)(10\ 12\ 11)$$

of order 3, it can be shown that M_{12} is the set $M_{11} \cup M_{11}\vartheta M_{11}$ and that M_{12} is transitive with the stabiliser of 12 being M_{11}. The argument to do this is exactly similar to that given in Proposition 24.3 and again in Proposition 24.4. This is left as an exercise. It follows that M_{12} is a group with 95 040 elements. This group is also simple, but this fact is best proved using more machinery from the general theory of permutation groups than we have developed. The interested reader should consult the book by Scott (1964) for details.

Remark 3 Similiar ideas may be used to construct another sequence of Mathieu groups, giving M_{20}, M_{21}, M_{22}, M_{23} and M_{24} with

$$|M_{24}| = 24 \cdot 23 \cdot 22 \cdot 21 \cdot 20 \cdot 48 = 244\,823\,040.$$

The basis for this sequence is the simple group $M_{21} = PSL(3,4)$ of order 20 160. Details of these constructions can also be found in Scott (1964). It may be shown that the groups M_{22}, M_{23} and M_{24} are all non-abelian simple groups. These three groups together with M_{11} and M_{12} are the five sporadic simple Mathieu groups.

Remark 4 We constructed M_{11} via a sequence of transitive groups: Q, M_9, M_{10} and M_{11} are transitive on sets with $8, 9, 10$ and 11 symbols, respectively. Furthermore Q is the stabiliser of a point in M_9, while M_9 is the stabiliser of a point in M_{10}, and M_{10} is the stabiliser of a point in M_{11}. In general, a group G is 1-transitive if it is transitive, and G is said to be k-transitive (for $k \geq 2$) if it is transitive and the stabiliser of a point is $(k-1)$-transitive on the points other than the point being stabilised.

In this terminology, the results in Propositions 24.3 and 24.4 can be generalised to give the following theorem.

Theorem 24.5 *Let G be a k-transitive permutation group on a set X with $k \geq 2$. Let X^+ be obtained from X by adjoining a point $*$. Suppose there is a permutation h on X^+, and an element $g \in G$, such that*

h interchanges $*$ with some x in X;
h fixes an element y in X;
g interchanges x and y;
$(gh)^3$ and h^2 are in G; and
$hG_xh = G_x$.

Then the group $\langle G, h\rangle$ is a $(k+1)$-transitive group on X^+ in which the stabiliser of $*$ is G.

This theorem was applied in Proposition 24.3 with $G = M_9$, X being the set $\{1, 2, \ldots, 9\}$, with $* = 10$, $h = (1\ 10)(4\ 5)(6\ 8)(7\ 9)$ and g being the permutatiion $(1\ 3)(4\ 9)(5\ 8)(6\ 7)$, and again in Proposition 24.4 with G being M_{10}, $X = \{1,2,\ldots,10\}$, $* = 11$, $h = (4\ 7)(5\ 8)(6\ 9)(10\ 11)$ and $g = (1\ 10)(4\ 5)(6\ 8)(7\ 9)$ However, as we have demonstrated, Theorem 24.5 is not essential to a construction of M_{11}.

We have seen that M_{11} is an example of a 4-transitive group. There are other obvious examples of highly transitive groups. The group $S(n)$ is n-transitive, and it may be shown that the alternating group $A(n)$ is $(n-2)$-transitive. These facts follow easily from the sequences $S(1) < S(2) < \cdots$ and $A(3) < A(4) < \cdots$, the kth term in each sequence being k-transitive. Apart from these examples, the only 4-transitive groups are M_{11} and M_{12}

(this is actually 5-transitive), M_{23} and M_{24} (this group is also 5-transitive). See Gorenstein (1982) for a deeper discussion of this fact which requires the classification of finite simple groups and so is well beyond the scope of this text.

Theorem 24.6 *The group M_{11} is a finite non-abelian simple group.*

Proof Since ρ_1 and ρ_2 do not commute, M_{11} is non-abelian. The proof of the simplicity of M_{11} can be divided into three steps. Let N be a non-trivial proper normal subgroup of $G = M_{11}$.

Step 1. We show that N is transitive on the set $\{1, \ldots, 11\}$. Since N is non-trivial, there is an element π, say, in N which moves an integer i, say, to $j \neq i$. Since G is transitive it contains an element ρ, say, such that $\rho(i) = 11$. Now let k be any integer in $\{1, 2, \ldots, 11\}$ not equal to i. We show that N is transitive by finding a permutation in N which takes i to k. Since $\pi(i) = j$, we may suppose that $j \neq k$. Since $\rho(i) = 11$, neither $\rho(j)$ nor $\rho(k)$ can equal 11, so there is a permutation σ in the transitive group M_{10} with $\sigma(\rho(j)) = \rho(k)$. Then

$$\rho^{-1}\sigma\rho(j) = \rho^{-1}\rho(k) = k \quad \text{and} \quad \rho^{-1}\sigma\rho(i) = \rho^{-1}\sigma(11) = i.$$

We have therefore found a permutation $\tau = \rho^{-1}\sigma\rho$ in G with $\tau(i) = i$ and $\tau(j) = k$. The conjugate $\tau\pi\tau^{-1}$ is therefore an element of N which moves i to the chosen k. This means that N is transitive, as required.

Step 2. We show that $|G : N| = 5$. Since N is transitive, 11 divides $|N|$ and so N contains a Sylow 11-subgroup of G, of order 11 since 11^2 does not divide $|G|$. Since N is a normal subgroup of G, it follows that N must contain all the Sylow 11-subgroups of G. The centraliser of an element σ of order 11 in $S(11)$ can easily be shown to consist of the powers of σ, and so has order 11. It follows that the centraliser of σ in M_{11} must also have order 11. Since the automorphism group of C_{11} has order 10 by Proposition 22.1, the normaliser–centraliser theorem, Proposition 10.26, shows that the normaliser of a Sylow 11-subgroup has order 11, 22, 55 or 110. Since the number of Sylow subgroups is the index of the normaliser, there are therefore either 720, 360, 144 or 72 Sylow 11-subgroups in M_{11}. The only one of these integers which is congruent to 1 mod 11 is 144, so there must be 144 Sylow 11-subgroups in M_{11}. It follows that the normal subgroup N contains all these 144 subgroups and so this subgroup contains at least $1 + 144(11 - 1) = 1441$ elements. The only proper divisors of 7920 greater than 1441 are 1584, 1980, 2640 and 3960. In any subgroup of order 2640 the number of Sylow 11-subgroups would divide $2640/11 = 240$, and similarly in any subgroup of order 1980 or 3960 the number of Sylow 11-subgroups would divide 180 or 360. Since the number of Sylow 11-subgroups is 144, we see that the only possibility is that $|N| = 1584$, so that the index $|G : N| = |G|/|N| = 5$, as required.

Step 3. Since $5 = |G : N|$ is not divisible by 3, any Sylow 3-subgroup of N (of order 9) is a Sylow 3-subgroup of G. Since all Sylow 3-subgroups of G are conjugate, and N is a normal subgroup of G, N contains all Sylow 3-subgroups of G. In particular, N contains the subgroup $\langle\pi_1, \pi_2\rangle$. Since N is a normal subgroup, $\pi_1\sigma\pi_2^2\sigma^{-1}$ is also an element of N. This element is

$$(1\ 2\ 3)(4\ 5\ 6)(7\ 8\ 9)(5\ 9\ 10)(2\ 6\ 4)(3\ 7\ 8) = (1\ 2\ 4\ 3\ 8)(5\ 7\ 9\ 10\ 6)$$

which has order 5, so that 5 divides the order of N. Since the order of N is 1584 it is not divisible by 5. This contradiction shows that G does not have a proper non-trivial normal subgroup, and so G is a non-abelian simple group. $\square$

Remark The group M_{11} is associated with two important combinatorial structures. The first of these is a well-known error-correcting code. Let X be the generator matrix

$$\begin{pmatrix} 1 & 0 & 0 & 0 & 0 & 0 & 1 & 1 & 1 & 1 & 1 \\ 0 & 1 & 0 & 0 & 0 & 0 & 0 & 1 & -1 & -1 & 1 \\ 0 & 0 & 1 & 0 & 0 & 0 & 1 & 0 & 1 & -1 & -1 \\ 0 & 0 & 0 & 1 & 0 & 0 & -1 & 1 & 0 & 1 & -1 \\ 0 & 0 & 0 & 0 & 1 & 0 & -1 & -1 & 1 & 0 & 1 \\ 0 & 0 & 0 & 0 & 0 & 1 & 1 & -1 & -1 & 1 & 0 \end{pmatrix}$$

over $\mathbf{Z}_3$. The corresponding code is the *ternary Golay code* $\mathcal{G}_{11}$. This code can be shown to have minimum distance 5. It is an example of a *perfect code*, in that every element of the 11-dimensional vector space over $\mathbf{Z}_3$ has distance 0, 1 or 2 from a unique codeword.

For any code C of length n, the automorphism group of C consists of the permutations in $S(n)$ which permute the n entries in the codewords so that the resulting n-vector is also a codeword. It can be shown that the automorphism group of the ternary Golay code is the Mathieu group M_{11}. In a similar way, the Mathieu group M_{24} is the automorphism group of the binary Golay code of length 24. This code is the starting point for the construction of the Leech lattice, associated with packing spheres in 24-dimensional space.

The other combinatorial idea is that of a *Steiner system*. A Steiner systen $S(r, s, t)$ is a collection of s-element subsets of a set S containing t elements, such that any selection of r elements from S lies in precisely one of the s-element subsets. Such configurations are of use when designing statistical experiments. It may be shown that there is a Steiner system $S(4, 5, 11)$ associated with the ternary Golay code, and the automorphism group of this Steiner system is also M_{11}. There is also a Steiner system $S(5,8,24)$ associated with the binary Golay code. The interested reader should consult the book by Anderson (1989) for further details on Steiner systems, sphere packings and the Leech lattice.

Summary for Chapter 24

In this section, we construct a group M_{11} by a series of *transitive extensions*. Starting from an explicitly constructed permutation group M_9 of order 72, we showed that M_9 is 2-transitive since the stabiliser of 1 is a group Q which is itself transitive on the remaining eight points. The group was then built up by explicit permutations to produce a transitive group M_{10} in which the stabiliser of 10 is M_9, and finally a transitive group M_{11} in which the stabiliser of 12 is M_{10}. The process can, in fact, be repeated once more to produce the 5-transitive group M_{12}. A similar method may be used to construct M_{24} by a sequence of one-point extensions. It was also proved that M_{11} is a non-abelian simple group.

The groups M_9, M_{10}, M_{11} and M_{12} were first constructed by Mathieu in 1861, with the sequence $M_{20}, M_{21}, M_{22}, M_{23}$ and M_{24} being introduced by him in 1873. The simplicity of the five Mathieu groups $M_{11}, M_{12}, M_{22}, M_{23}$ and M_{24} was established by Miller in 1900.

25

The classification of finite simple groups

One of the major intellectual achievements of all time has been the classification of the finite simple groups. As we saw in the chapter on the Jordan–Hölder Theorem, these are the 'atoms' of finite group theory. In this chapter, we present a survey of the groups occurring in the list of finite simple groups.

We first give an outline of the types of groups which appear in the classification. These are:

(1) the abelian simple groups: these are cyclic groups of prime order;
(2) the alternating groups $A(n)$ for $n \geq 5$;
(3) various families of groups of *Lie type*. The easiest examples of groups in this class are the groups $PSL(2, q)$ discussed in Chapter 23;
(4) the sporadic groups: a set of 26 simple groups which are not accounted for in the previous three categories. The easiest example of a sporadic group is the group M_{11} discussed in Chapter 24.

The first two lists of groups, the cyclic groups of prime order and the alternating groups, are already familiar to us. As a preliminary to an explanation of the groups of Lie type, we shall first discuss the *classical groups*. There are several families of classical groups, all associated with matrix groups over finite fields. These families will be discussed in Section 25.1. In Section 25.2, we shall give a very brief explanation of the way in which the classical groups were unified by introducing groups of Lie type. Finally, in Section 25.3, we shall say something about the 26 sporadic groups. It will not be possible to give any indication of how it was shown that the list of groups we describe is the complete list of finite simple groups. A deeper account of the classification can be obtained by consulting the book by Gorenstein (1982). Very few of the assertions made in the remainder of this chapter will be proved. The intention is simply to acquaint the reader with these fundamental objects, and to give an impression of the complexity of the task of establishing the classification of the finite simple groups.

25.1 The classical groups

As a starting point for a discussion of the groups of Lie type, we first consider the *classical groups*. There are several families of classical groups, each of which is associated with a matrix group over a finite field. We have already discussed the simplest of these.

25.1.1 The projective special linear groups

The first family of classical groups are the groups $PSL(n, q)$, where q is a power of a prime p. These were defined in Chapter 23, and, as mentioned there, these groups are simple if $n > 2$ or if $n = 2$ and $q \geq 4$. There are some 'exceptional' isomorphisms between these simple groups and the alternating groups. As we have already mentioned, $PSL(2, 4)$ and $PSL(2, 5)$ are both isomorphic to the group $A(5)$; also $PSL(4, 2)$ is isomorphic to $A(8)$ and $PSL(2, 9) \cong A(6)$. In addition $PSL(2, 7) \cong PSL(3, 2)$. With these exceptions, it may be shown that the simple groups in the list $PSL(n, q)$ are distinct from one another and from the alternating groups.

The groups $PSL(n, q)$, and their simplicity, were discussed by Jordan in the case when q is a prime in his famous *Traité des Substitions* of 1870. The generalisation to the case of matrices over an arbitrary finite field was carried out by Dickson during the 1890s.

25.1.2 The unitary groups

The other classical groups are obtained from $GL(n, q)$ by considering those matrices which *preserve* certain *forms*. To illustrate this, we first consider the *unitary groups*.

Let q be a power of a prime p, and let F be the field with q^2 elements. The map $x \mapsto x^q$ is an automorphism of F. Since the multiplicative group of F is of order $q^2 - 1$, it is easily seen that this automorphism has order two.

Definition 25.1 *A* sequilinear form *on a vector space V of dimension n over the field F with q^2 elements is a map $f : V \times V \to F$ satisfying the following conditions. For all $u, v, w \in V$ and all $\lambda \in F$:*

(1) $f(u + v, w) = f(u, w) + f(v, w)$;
(2) $f(\lambda u, v) = \lambda f(u, v)$;
(3) $f(u, v + w) = f(u, v) + f(u, w)$; *and*

(4) $f(u, \lambda v) = \lambda^q f(u, v)$.

The form is said to be hermitian *if also*

(5) $f(u, v) = f(v, u)^q$.

Definition 25.2 *A matrix transformation X on V is said to* preserve a *hermitian form f if $f(Xu, Xv) = f(u, v)$ for all u and v in V.*

Remark A form f is non-degenerate if, whenever $f(u, v) = 0$ for all v, then $u = 0$. It can be shown that the group of matrices preserving a hermitian form is independent of the form (up to conjugacy), provided that the form is non-degenerate. In fact, it can be shown that any two groups preserving non-degenerate forms are conjugate subgroups of the group $GL(n, q^2)$.

We may translate these definitions into matrix terms. In order to do this, for any matrix X over F, we introduce the notation $\bar{X}$ to denote the matrix obtained from X by replacing each entry of X with its image under the automorphism $x \mapsto x^q$. Now, in order to study a form on $V = V(n, q^2)$, we choose a fixed basis $v_1, v_2, \ldots, v_n$ for V, and consider coordinate (column) vectors with respect to this basis. Let A be the $n \times n$ matrix whose (i, j)th entry is $f(v_j, v_i)$. It follows from the definition that for all column vectors u and v in V,

$$f(u, v) = \bar{v}^{\mathrm{T}} Au.$$

We refer to A as the matrix of the form f. It may be shown that if f is a non-degenerate hermitian form, then there is a basis for the vector space $V(n, q^2)$ such that the matrix of f is the identity matrix I, so that

$$f(Uu, Uv) = (\overline{Uv})^{\mathrm{T}} IUu = \bar{v}^{\mathrm{T}} \bar{U}^{\mathrm{T}} Uu.$$

Thus, if a matrix U preserves this non-degenerate hermitian form, we see that, for all u and v,

$$\bar{v}^{\mathrm{T}} \bar{U}^{\mathrm{T}} Uu = f(Uu, Uv) = f(u, v) = \bar{v}^{\mathrm{T}} Iu,$$

from which it follows that $\bar{U}^{\mathrm{T}} U = I$. Such a matrix U is said to be *unitary*. A unitary matrix U therefore satisfies the equation $U^{-1} = \bar{U}^{\mathrm{T}}$. Hence if δ is the determinant of a unitary matrix, $\delta^{-1} = \delta^q$, so that $\delta^{1+q} = 1$.

Example 25.3 Consider the 2×2 unitary matrices over $GF(2^2)$. Let ω be a generator of the multiplicative group in $GF(2^2)$, so that $\omega^3 = 1$ and $\omega \neq 1$. Let

$$U = \begin{pmatrix} a & b \\ c & d \end{pmatrix}$$

be a unitary matrix with determinant δ. Since U is unitary,

$$\delta^{-1}\begin{pmatrix} d & -b \\ -c & a \end{pmatrix} = \begin{pmatrix} a^2 & c^2 \\ b^2 & d^2 \end{pmatrix}.$$

Thus $d = \delta a^2$, so that $d^2 = \delta^2 a = \delta^{-1}a$. It follows that if a is non-zero, so is d, and in that case, $ad = \delta$. Thus, one of b and c must be zero, and since $-b = \delta c^2$, we deduce that both b and c are 0. If, however, $b \neq 0$ it follows that $a = 0$ so that $d = 0$. It follows from this discussion that the unitary 2×2 matrices over $GF(2^2)$ are the following 18 matrices:

$$\begin{pmatrix} \lambda & 0 \\ 0 & \mu \end{pmatrix} \text{ where } \lambda, \mu \neq 0, \text{ and } \begin{pmatrix} 0 & \lambda \\ \mu & 0 \end{pmatrix} \text{ where } \lambda, \mu \neq 0.$$

The six unitary matrices of determinant 1 are

$$\begin{pmatrix} 1 & 0 \\ 0 & 1 \end{pmatrix}, \begin{pmatrix} \omega & 0 \\ 0 & \omega^2 \end{pmatrix}, \begin{pmatrix} \omega^2 & 0 \\ 0 & \omega \end{pmatrix}, \begin{pmatrix} 0 & 1 \\ 1 & 0 \end{pmatrix}, \begin{pmatrix} 0 & \omega \\ \omega^2 & 0 \end{pmatrix} \text{ and } \begin{pmatrix} 0 & \omega^2 \\ \omega & 0 \end{pmatrix}$$

and these form a subgroup isomorphic to the group $SL(2,2)$.

The unitary matrices of determinant 1 form a group since they are the intersection of the unitary group with $SL(n,q^2)$. The factor group of this group by the subgroup of scalar matrices of determinant 1 will be denoted by $U_n(q)$. We have just seen that $U_2(2)$ is isomorphic to $PSL(2,2)$. In fact, it can be shown that for all prime powers q, the group $U_2(q)$ is isomorphic to $PSL(2,q)$.

For general values of n the order of $U_n(q)$ can be shown to be

$$q^{n(n-1)/2}(q^n - (-1)^n)(q^{n-1} - (-1)^{n-1}) \ldots (q^3 + 1)(q^2 - 1)/d,$$

where d is the greatest common divisor of $q+1$ and n. Thus $U_3(2)$ has 72 elements. This group is in fact a semidirect product of a group isomorphic to $C_3 \times C_3$ by a quaternion group of order 8 (in fact $U_3(2)$ is isomorphic to the group M_9 discussed in Chapter 24). In all other cases when $n > 2, U_n(q)$ is a simple group and is not isomorphic to any of the other simple groups we have mentioned. Thus, the first new example of a simple group in this series is the group $U_3(3)$ of order 6048.

25.1.3 The symplectic groups

Next we consider the *symplectic groups.* The starting point here is the idea of an alternating form.

Definition 25.4 *A* bilinear form *on a vector space V of dimension n over the field F is a map $f : V \times V \to F$ satisfying the following conditions. For all $u, v, w \in V$ and all $\lambda \in F$:*

(1) $f(u + v, w) = f(u, w) + f(v, w)$;
(2) $f(\lambda u, v) = \lambda f(u, v)$;
(3) $f(u, v + w) = f(u, v) + f(u, w)$; *and*
(4) $f(u, \lambda v) = \lambda f(u, v)$.

A bilinear form is alternating *if for all u, v in V,*

$$f(u, v) = -f(v, u).$$

It is an obvious consequence of the definition that if f is an alternating form and F does not have characterstic 2, then for all v in $V, f(v, v) = 0$. A matrix which preserves a non-degenerate alternating form is said to be *symplectic.*

Remark It can be also shown that the group of matrices preserving a non-degenerate alternating form is independent of the form (up to conjugacy). Also, two groups preserving non-degenerate forms are conjugate subgroups in $\mathrm{GL}(n, q)$. If f is a non-degenerate alternating form, it can be shown that the vector space V must have even dimension $n = 2m$, say. Then we can find a basis $u_1, u_2, \ldots, u_{2m}$ for V such that

$$f(u_i, u_{m+i}) = 1, \; f(u_{m+i}, u_i) = -1 \quad \text{for } 1 \le i \le m$$

and

$$f(u_i, u_j) = 0 \quad \text{if } j \ne i + m \text{ or } i \ne j + m.$$

Example 25.5 Consider the case when $n = 1$. Let A be the matrix

$$A = \begin{pmatrix} 0 & -1 \\ 1 & 0 \end{pmatrix}$$

and f be the bilinear form defined by $f(u, v) = v^{\mathrm{T}} Au$. Thus if

$$u = \begin{pmatrix} \lambda_1 \\ \lambda_2 \end{pmatrix} \text{ and } v = \begin{pmatrix} \mu_1 \\ \mu_2 \end{pmatrix},$$

the values of f are given by

$$f(u,v) = \lambda_1\mu_2 - \lambda_2\mu_1.$$

It follows that f is alternating because

$$f(u,v) = \lambda_1\mu_2 - \lambda_2\mu_1 = -(\lambda_2\mu_1 - \lambda_1\mu_2) = -f(v,u).$$

The matrix of this form is therefore A. Suppose that the matrix

$$\begin{pmatrix} a & b \\ c & d \end{pmatrix}$$

preserves this form. Then

$$\begin{pmatrix} a & b \\ c & d \end{pmatrix}^{\mathrm{T}} \begin{pmatrix} 0 & -1 \\ 1 & 0 \end{pmatrix} \begin{pmatrix} a & b \\ c & d \end{pmatrix} = \begin{pmatrix} 0 & -1 \\ 1 & 0 \end{pmatrix}.$$

We therefore obtain:

$$\begin{pmatrix} a & c \\ b & d \end{pmatrix} \begin{pmatrix} -c & -d \\ a & b \end{pmatrix} = \begin{pmatrix} 0 & -1 \\ 1 & 0 \end{pmatrix},$$

so that

$$-ac + ac = 0,\ ad - bc = 1,\ -ad + bc = -1 \text{ and } -bd + bd = 0.$$

Thus we see that the symplectic 2×2 matrices are precisely the matrices of determinant 1.

In fact, for any even $n = 2m$, symplectic matrices always have determinant 1. The quotient group of the group $Sp_n(q)$ of symplectic matrices by the scalar symplectic matrices is the group $S_n(q)$ (or $PSp_n(q)$). It can be shown that the group $S_{2m}(q)$ has order

$$q^{m^2}(q^{2m} - 1)(q^{2m-2} - 1)\ldots(q^4 - 1)(q^2 - 1)/d,$$

where $d = 2$ if q is odd and $d = 1$ if q is even.

We have seen that $S_2(q)$ is equal to $PSL(2,q)$. For $n \geq 2$, the group $S_n(q)$ is simple, except that $S_4(2)$ is isomorphic to the symmetric group $S(6)$. It may also be shown that $S_4(3)$ is isomorphic to $U_4(2)$. Hence, the first new group in the list of finite simple groups arising from the sympletic groups is $S_4(4)$ of order $4^4 \times 255 \times 15 = 979\,200$. Also, with the exception of $S_4(3)$, for $n \geq 2$ the simple groups in the list $S_{2m}(q)$ are not isomorphic to any other simple groups in the list of groups we have constructed so far.

25.1.4 The orthogonal groups

Next, we consider the orthogonal groups. The situation here is much more complicated than for our previous families of groups. This arises from the fact that for even n there are two forms giving rise to non-isomorphic orthogonal groups.

We first collect together some basic definitions. A bilinear form f is *symmetric* if, for all u and v in V, $f(u,v) = f(v,u)$. Associated with a symmetric bilinear form is a *quadratic form* F defined by

$$F(\lambda u + \mu v) = \lambda^2 F(u) + \lambda\mu f(u,v) + \mu^2 F(v).$$

The *kernel of* f is the set of u such that $f(u,v) = 0$ for all $v \in V$. The *kernel of* F is the set of u in the kernel of f such that $F(u) = 0$. We say that f or F is *non-singular* when its kernel is $\{0\}$.

We shall first discuss the easiest example of the construction. This occurs for odd degree matrices when q is also odd. The advantage of taking q to be odd is that f and F determine each other in this case. The advantage of requiring n to be odd is that there is only one equivalence class of non-singular symmetric bilinear forms. The orthogonal group, $GO(n,q)$, is the set of matrices which preserve the symmetric bilinear form f. As usual, the subgroup $SO(n,q)$ is obtained by taking elements of determinant 1. Also $PSO(n,q)$ is defined to be the factor group of $SO(n,q)$ by the scalar matrices it contains. However, $PSO(n,q)$ is not simple, but has a simple subgroup of index two, and it is this group which is defined to be the group $O_n(q)$. The appropriate subgroup of $SO(n,q)$ is its commutator subgroup which has index at most 2. This is denoted $\Omega_n(q)$ and may also be defined as follows. Let r be any element of the underlying vector space $V(n,q)$. Define the *reflection in* r to be the element of $GO(n,q)$ determined by

$$x \mapsto x - \frac{f(x,r)}{F(r)} \cdot r.$$

Each element in $GO(n,q)$ can be written as a product of suitable reflections. Now define $\Omega_n(q)$ to be the set of all g in $SO(n,q)$ such that, whenever g is expressed as a product of reflections in vectors $r_1, r_2, \ldots, r_s$, then the element $f(r_1,r_1)f(r_2,r_2)\ldots f(r_n,r_n)$ is a square in the field $GF(q)$. Then $O_n(q)$ is defined to be the factor group of $\Omega_n(q)$ by the scalar matrices which it contains. This is a group of order

$$q^{m^2}(q^{2m}-1)(q^{2m-2}-1)\ldots(q^4-1)(q^2-1)/2,$$

where $n = 2m+1$. The group $O_3(q)$ is isomorphic to $PSL(2,q)$ and the group $O_5(q)$ is isomorphic to the symplectic group $S_4(q)$. For $m \geq 3$, $O_{2m+1}(q)$ is not isomorphic to any other simple group.

When $n = 2m$ is even and $n \geq 4$, but q is still odd, it turns out that there are two inequivalent non-singular quadratic forms on $V(n,q)$. These differ by having different *Witt index*; this is the greatest dimension of any subspace U for which $F(u) = 0$ for all $u \in U$. It turns out that the *plus type* of quadratic form has Witt index m, whereas the *minus type* has Witt index $m-1$. These give two different groups of orthogonal matrices denoted by $O_n^+(q)$ and $O_n^-(q)$. We define subgroups $\Omega_n^+(q)$ and $\Omega_n^-(q)$ of $SO^+(n,q)$ and $SO^-(q,n)$, respectively. These are each of index two. Taking the quotient group by the subgroup of scalar matrices gives the groups

$$P\Omega_n^+(q) = O_n^+(q) \text{ and } P\Omega_n^-(q) = O_n^-(q).$$

The formulae for their orders are

$$|O_n^+(q)| = q^{m(m-1)}(q^n - 1)(q^{2m-2} - 1)\ldots(q^4 - 1)(q^2 - 1)/d,$$

where d is the greatest common divisor of 4 and $q^n - 1$, and

$$|O_n^-(q)| = q^{m(m-1)}(q^n + 1)(q^{2m-2} - 1)\ldots(q^4 - 1)(q^2 - 1)/d,$$

where d is the greatest common divisor of 4 and $q^n + 1$. As usual, there are some isomorphisms with other classical groups for small n. In this case these are:

$$O_4^+(q) = PSL(2,q) \times PSL(2,q), \quad O_4^-(q) = PSL(2,q^2),$$
$$O_6^+(q) = PSL(4,q) \text{ and } O_6^-(q) = U_4(q).$$

In particular, $O_4^+(q)$ is never simple. Otherwise, for even $n, O_n^+(q)$ and $O_n^-(q)$ are simple groups not in our previously constructed lists.

Finally, we discuss briefy the case when q is even. If $n = 2m+1$, the group $\mathrm{SO}(n,q)$ is isomorphic to the symplectic group $S_{2m}(q)$.

When n is even, it is still possible to define subgroups $O_n^+(q)$ and $O_n^-(q)$ of $\mathrm{SO}^+(n,q)$ and $\mathrm{SO}^-(q,n)$, respectively, each of index two. The formulae for the orders of the corresponding groups coincide with those for odd q, as do the 'exceptional' isomorphisms for small degrees.

It is of interest to point out that for any q and $m \geq 3$, the simple groups $O_{2m+1}(q)$ and $S_{2m}(q)$ have the same order, but are non-isomorphic.

This completes the list of the classical groups. The reader who wishes to find proofs for the assertions in this section should consult the book by Artin (1957) or that by Dieudonné (1948).

The classical groups were known to Dickson at the turn of the century. In most cases, the existence and simplicity of the classical groups over the base field when $q = p$ had been discussed by Jordan in his *Traité des Substitutions* of 1870, and the extension to arbitrary finite fields was carried out by Dickson.

25.2 Groups of Lie type

The work of Chevalley (1955) and Steinberg (1959) unified the list of classical groups and led to the discovery of other infinite families of finite simple groups. Unfortunately this work is much too technical to be included here. We can only give a very general introduction to this subject and refer the reader to the book by Carter (1985) for details.

The basic idea arises from the classification of complex simple Lie algebras. A *complex Lie algebra* is a complex vector space with a non-associative multiplication $[xy]$ satisfying the axioms:

$$[x(y+z)] = [xy] + [xz] \quad \text{and} \quad [(x+y)z] = [xz] + [yz];$$

$$[[xy]z] + [[zx]y] + [[yz]x] = 0 \quad \text{and} \quad [(\lambda x)y] = \lambda[xy].$$

The complete list of simple (i.e. with no proper ideals) non-abelian (i.e there exist x and y such that $[xy] \neq 0$) Lie algebras was determined by Cartan in 1894 in an article reprinted in his complete works of 1952. The resulting classification provivides a list of algebras known as

$$A_n \quad (\text{for } n \geq 1), \; B_n \quad (\text{for } n \geq 2), \; C_n \quad (\text{for } n \geq 3), \; D_n \quad (\text{for } n \geq 4),$$

$$G_2, \; F_4, \; E_6, \; E_7 \text{ and } E_8.$$

It turns out that it is possible to associate an infinite family of finite groups with each algebra in this list (one group for each finite field). The corresponding list of groups is the list of *adjoint Chevalley groups.* With a few small exceptions, these groups are finite simple groups. In fact, the groups associated with the algebras of type A_n are precisely the groups $PSL(n+1,q)$, those of type B_n are $O_{2n+1}(q)$, the groups of type C_n are the symplectic groups $S_{2n}(q)$ and those corresponding to type D_n are the orthogonal groups $O^+_{2n}(q)$. This unification was made by Chevalley, who at the same time discovered the families of non-classical simple groups of types F_4, E_6, E_7 and E_8. The groups of type G_2 were known earlier.

The next step was the work of Steinberg, who produced the *twisted groups*

$$^2A_n, \; ^2D_n, \; ^3D_4 \text{ and } ^2E_6.$$

These arise because there is a graph, known as the Dynkin diagram, associated with each simple Lie algebra. In the case of simple Lie algebras of types A_n, D_n and E_6 the corresponding Dynkin diagram has an automorphism of order 2, while the diagram of type D_4 admits an automorphism of order 3. It turns out that in these cases of graph automorphisms of order t, say, there is a twisted Chevalley group, $t_n^X(q, q^t)$, more usually written as $t_n^X(q)$. This is defined to be the subset of elements of $X_n(q)$ that are fixed by the quotient of the graph automorphism and the field automorphism induced by the Frobenius automorphism $x \mapsto x^q$ of $GF(q^t)$.

It turns out that groups of type ${}^2A_n$ correspond to the unitary groups $U_{n+1}(q)$, those of type ${}^2D_n$ to the orthogonal groups $O_{2n}^-(q)$, while those of types ${}^3D_4$ and ${}^2E_6$ were newly discovered. Finally, three families admitted *exceptional automorphisms* involving the graph and the field; these occur in cases B_2, G_2 and F_4. The corresponding families occur only when q is a power of a special prime, giving the Suzuki groups (where q is even), Ree groups of characteristic three and Ree groups of characteristic two. The first group in each of these three families is not, in fact, simple, but all later members of the families are simple. The exceptional groups are:

${}^2B_2(2)$, a non-abelian group of order 20;

${}^2G_2(2)$, an extension of $PSL(2,8)$ by a group of order 3;

${}^2F_4(2)$, an extension of a group T (a simple group known as the Tits group) by a group of order 2.

25.3 The sporadic groups

The final ingredient in the classification is the set of 26 sporadic groups, none of which occurs in any of the infinite families. Five of these were discovered by Mathieu in 1861 and 1873. It was not until 1965 that the next group J_1 in the list was discovered by Janko. The majority of these groups were discovered in the 1960s and 1970s. One important technique in finding them was to attempt to calculate the list of finite simple groups for which the centraliser of an element of order 2 has a given structure. Thus, for example, Janko proved in 1966 that if G is a finite simple group with abelian Sylow 2-subgroup of order 8 and the centraliser of an element of order 2 is isomorphic to $C_2 \times PSL(2,5)$, then G is isomorphic to J_1 of order 175 560. In fact, 11 of the 26 sporadic groups can be characterised in this way by the structure of the centraliser of an element of order 2.

Other sporadic groups were discovered using a variety of methods including the classification of the rank 3 permutation groups (these are transitive permutation groups in which the stabiliser of a point has three orbits on the underlying set), and the groups discovered by Conway associated with the Leech lattice.

The final finite simple group in the list was the Monster which was shown to exist by Griess in 1981. This is the last group in the list of sporadic groups in Table 25.1. For each group in the list, we give only its name in the ATLAS of finite simple groups, as well as its order and year

Table 25.1 The sporadic finite simple groups

Name	Order	Discovered
M_{11}	$2^4 3^2 5.11$	1861
M_{12}	$2^6 3^3 5.11$	1861
M_{22}	$2^7 3^2 5.7.11$	1873
M_{23}	$2^7 3^2 5.7.11.23$	1873
M_{24}	$2^{10} 3^3 5.7.11.23$	1873
J_1	$2^3 3.5.7.11.19$	1966
J_2	$2^7 3^3 5^2 7$	1968
J_3	$2^7 3^5 5.17.19$	1968
J_4	$2^{21} 3^3 5.7.11^3 23.29.31.37.43$	1976
HS	$2^9 3^2 5^3 7.11$	1968
McL	$2^7 3^6 5^3 .7.11$	1969
Suz	$2^{13} 3^7 5^2 7.11.13$	1969
Ru	$2^{14} 3^3 5^3 7.13.29$	1973
He	$2^{10} 3^3 5^2 7^3 .17$	1969
Ly	$2^8 3^7 5^6 7.11.31.37.67$	1972
O'N	$2^9 3^4 5.7^3 11.19.31$	1976
Co3	$2^{10} 3^7 5^3 7.11.23$	1969
Co2	$2^{18} 3^6 5^3 7.11.23$	1969
Co1	$2^{21} 3^9 5^4 7^2 11.13.23$	1969
Fi(22)	$2^{17} 3^9 5^2 7.11.13$	1971
Fi(23)	$2^{18} 3^{13} 5^2 7.11.13.17.23$	1971
Fi(24)′	$2^{21} 3^{16} 5^2 7^3 11.13.17.23.29$	1971
HN	$2^{14} 3^6 5^6 7.11.19$	1976
Th	$2^{15} 3^{10} 5^3 7^2 13.19.31$	1976
BM	$2^{41} 3^{13} 5^6 7^2 11.13.17.19.23.31.47$	1976
M	$2^{46} 3^{20} 5^9 7^6 11^2 13^3 17.19.23.29.31.41.47.59.71$	1981

of discovery. The reader who wishes to obtain more details should either consult the ATLAS of finite simple groups (Conway *et al.* 1985) the account in Gorenstein (1982) or the book by Aschbacher (1994).

Summary for Chapter 25

In this chapter, we discuss the classification of the finite simple groups. It is now known that any finite simple group is isomorphic to a group in the following list:

(1) the cyclic groups of prime order;

(2) the alternating groups $A(n)$ for $n \geq 5$;
(3) various families of groups of Lie type;
(4) the sporadic groups.

Our objective in this chapter is to offer an introduction to the groups in this list. The main discussion is about the classical groups. These are obtained from groups of matrices over finite fields satisfying certain restrictions. The results on classical groups were unified by Chevalley (1955) and Steinberg (1959), when these groups were seen as groups of Lie type. At this time, the search for sporadic groups also gathered momentum. Not surprisingly, these were mostly discovered by *ad hoc* methods, although there were times at which it appeared that some pattern was being established. There is still a great deal of mystery surrounding the Monster M and its connection with number theory and physics. In some ways this behaves as a 'universal' simple group in that the majority of the other 25 sporadic groups occur as subgroups of M.

The list of finite simple groups is merely one aspect of the classification theorem. One needs to prove that any finite simple group is isomorphic to one of the groups in this list. The reader who wishes to understand how this was proved should consult the book by Gorenstein (1982).

Gorenstein estimates that the full details of the classification occupy about 15 000 pages in research journals and that this work involved about 400 research mathematicians from many different countries. Apart from Gorenstein himself, one should mention that Thompson and Aschbacher are also among the major contributors to the classification. Proofs of this complexity are almost certain to contain errors of detail. However, it is felt that the evidence for the correctness of the classification is now overwhelming. In particular, the only errors which will arise are minor ones in the methods of proof rather than in the statement of the outcome.

It is not unreasonable, therefore, to claim this classification as one of the greatest of all intellectual achievements. It is a great pity that this tribute to human ingenuity has passed mostly unnoticed outside the mathematical community. One of the objectives in writing this book was to attempt to take a small step on the way to rectifying this injustice.

A

Prerequisites from number theory and linear algebra

In this Appendix, we shall mention some of the facts from other areas of mathematics used in the book.

A.1 Number theory

There are two main results required from number theory. The first of these is known as the division algorithm.

Theorem A.1 *Given two positive integers a and b, there are integers q and r with $0 \leq r \leq b-1$ such that $a = bq + r$.*

Proof The proof is a simple application of the well-ordering principle that every non-empty subset of non-negative integers has a least element. Define a set S to consist of those non-negative integers which have the form $a - bn$ ($n \in \mathbf{Z}$). Since a is in S (take $n = 0$), S is non-empty and so has a least element, r, say. Then $r = a - bq$ for some q. If r were greater than $b - 1$, then $r - b$ would be a non-negative integer less than r, and since $r - b = a - b(q + 1)$, we see that $r - b$ would also be in S, contrary to the choice of r. It follows that $0 \leq r \leq b - 1$, as required. □

The other basic result is the following.

Theorem A.2 *Given positive integers a and b, the smallest positive integer d of the form $as + bt$ satisfies the conditions:*

(1) *d divides a and d divides b; and*
(2) *if e divides both a and b, then e divides d.*

Proof The proof is another application of the well-ordering principle. Let

$$D = \{as + bt : s \text{ and } t \text{ are integers and } as + bt > 0\}.$$

Since $a = (1 \times a + 0 \times b)$ is in D, the set is non-empty and so has a least element d, say. It follows that d is of the form $as + bt$ for some s and t. We must show that d satisfies conditions (1) and (2).

To check (1), use Theorem A.1 to write $a = qd + r$ with $0 \leq r \leq d - 1$. If r is non-zero, we see that

$$r = a - qd = a - q(as + bt) = a(1 - qs) + b(-qt).$$

It follows by definition that r is in D, contrary to the minimality of d. We deduce that r is zero, so that d divides a. The proof that d divides b is similar.

To check (2), suppose that e divides a and b. Then $a = eu$ and $b = ev$ for some integers u and v. Thus

$$d = as + bt = eus + evt = e(us + vt),$$

so that e divides d. □

Definition A.3 *For integers a and b, the integer d constructed as in Theorem A.2 is known as the* greatest common divisor *of a and b.*

There is an algorithm to calculate the greatest common divisor of a and b known as the *Euclidean algorithm.* It is obtained by repeated application of the division algorithm. Write

$$a = bq + r \quad \text{with } 0 \leq r < b.$$

Repeat with a replaced by b and b replaced by r, to obtain

$$b = rq_1 + r_1 \quad \text{with } 0 \leq r_1 < r.$$

Repeat again with a replaced by r and b by r_1:

$$r = r_1 q_2 + r_2 \quad \text{with } 0 \leq r_2 < r - 1.$$

Since the sequence of remainders is a strictly decreasing set of non-negative integers

$$r > r_1 > r_2 > \cdots,$$

it must terminate in zero after a finite number k, say, of steps. The last non-zero remainder r_{k-1} is the greatest common divisor of a and b. This fact is proved by showing that r_{k-1} satisfies conditions (1) and (2) of Theorem A.2. The interested reader is referred to Humphreys and Prest (1989) for details.

A.2 Linear algebra and determinants

Unlike the results used from number theory, we shall not attempt to prove any of the required results from linear algebra. There are many books on the subject, all of which will give more details on the following basic facts.

Definition A.4 *Let $u_1, u_2, \ldots, u_r$ be n-vectors with entries from a field F. These vectors are* linearly independent *if the only elements $\lambda_1, \lambda_2, \ldots, \lambda_r$ in F which satisfy the equation*

$$\lambda_1 u_1 + \lambda_2 u_2 + \cdots + \lambda_r u_r = 0$$

are the elements

$$\lambda_1 = \lambda_2 = \cdots = \lambda_r = 0.$$

There is an associated concept of a spanning set.

Definition A.5 *Let $u_1, u_2, \ldots, u_r$ be n-vectors with entries from a field F. These vectors are a* spanning set *if every n-vector can be written as a linear combination of $u_1, u_2, \ldots, u_r$.*

Definition A.6 *A linearly independent set which is also a spanning set is a* basis.

Theorem A.7 *Let $u_1, u_2, \ldots, u_r$ be a linearly independent set of vectors of length n. Then $r \leq n$. If $u_1, u_2, \ldots, u_s$ is a spanning set of vectors of length n, then $s \geq n$. Any linearly independent set $u_1, u_2, \ldots, u_n$ of vectors of length n is a spanning set.* □

One consequence of this fact which is used in Chapter 23 is the following.

Corollary A.8 *Let u and v be non-zero vectors of length two such that v is not a multiple of u. Then if w is any other vector, there exist field elements a and b such that $w = av + bv$.*

Proof The hypothesis ensures that u and v are linearly independent, so the result follows directly from Theorem A.7. □

We mention some basic properties of determinants. The first two of these properties are used repeatedly.

Theorem A.9 *Let A and B be $n \times n$ matrices. Then:*

(1) $\det(AB) = \det(A)\det(B)$;

(2) *the matrix A has an inverse if and only if $\det(A)$ is non-zero.* □

The next set of properties are used in Chapter 23.

Theorem A.10 *Let A be an $n \times n$ matrix. Then:*

(1) *if all the entries in a column of A are zero, then the determinant of A is zero;*

(2) *if B is obtained by interchanging any two columns of A, then*

$$\det(A) = -\det(B);$$

(3) *if B is obtained from A by multiplying each entry in a fixed column of A by λ, then* $\det(B) = \lambda\det(A)$;

(4) *if B is obtained by adding any multiple of the i-th column of A to its j-th column, then* $\det(A) = \det(B)$. □

B

Groups of order < 32

Recall that for any integer n, there is always a group with n elements, namely, the cyclic group C_n. For every even integer $2n$ greater than 2, there are at least two non-isomorphic groups with $2n$ elements, the cyclic group C_{2n} and the dihedral group $D(n)$. For every integer of the form $4n$ greater than 4, there are at least three non-isomorphic groups with $4n$ elements: the cyclic group C_{4n}, the dihedral group $D(2n)$ and the generalised quaternion group Q_{4n}, as defined in Proposition 22.7.

1. The prime integers between 1 and 31 are $2, 3, 5, 7, 11, 13, 17, 19, 23, 29$ and 31. For each of these orders, there is precisely one group, the cyclic group, of that order by Proposition 5.19. In addition, there is precisely one group of order 15 by Example 11.12.

2. The numbers which are squares of primes up to 31 are 4, 9 and 25. By Corollary 10.23 there are two isomorphism types of groups of order p^2, the cyclic group and the elementary abelian p-group isomorphic to $C_p \times C_p$.

3. When p is an odd prime, there are precisely two isomorphism types of group of order $2p$, the cyclic group and the dihedral group $D(p)$ (see section 1 of Chapter 12). The integers in our range of the form $2p$ are $6, 10, 14, 22$ and 26.

The integers left for further discussion are $8, 12, 16, 18, 20, 21, 24, 27, 28$ and 30. We consider these in turn.

4. There are three abelian groups with 8 elements; C_8, $C_4 \times C_2$ and $C_2 \times C_2 \times C_2$. The non-abelian groups of order 8 are the dihedral group $D(4)$ and the quaternion group Q_8. Both of these groups have centres of order two, and in addition the derived groups coincide with the centres.

5. An abelian group of order 12 is either cyclic or isomorphic to $C_2 \times C_6$. For the non-abelian groups of order 12 the Sylow 3-subgroup is cyclic, and the main properties of these groups are shown in Table B.1.

Table B.1 Non-abelian groups of order 12

Group	G'	$Z(G)$	Sylow 2-subgroup
$D(6)$	C_3	C_2	$C_2 \times C_2$
$A(4)$	$C_2 \times C_2$	$\{1\}$	$C_2 \times C_2$
Q_{12}	C_3	C_2	C_4

6. The abelian groups of order 16 are C_{16}, $C_8 \times C_2$, $C_4 \times C_4$, $C_4 \times C_2 \times C_2$ and $C_2 \times C_2 \times C_2 \times C_2$. The non-abelian groups of order 16 are as shown below.

Table B.2 Non-abelian groups of order 16

Group	G'	$Z(G)$
$C_2 \times D(4)$	C_2	$C_2 \times C_2$
$C_2 \times Q_8$	C_2	$C_2 \times C_2$
$D(8)$	C_4	C_2
Q_{16}	C_4	C_2
$\langle x, y : x^8 = 1 = y^2, x^y = x^3 \rangle$	$\langle x^2 \rangle$	$\langle x^4 \rangle$
$\langle x, y : x^8 = 1 = y^2, x^y = x^5 \rangle$	$\langle x^4 \rangle$	$\langle x^2 \rangle$
$\langle x, y, z : x^4 = 1 = y^2 = z^2, x \text{ central}, z^y = zx^2 \rangle$	$\langle x^2 \rangle$	$\langle x \rangle$
$\langle x, y : x^4 = 1 = y^4, x^y = x^3 \rangle$	$\langle x^2 \rangle$	$\langle x^2, y^2 \rangle$
$\langle x, y, z : x^4 = 1 = y^2 = z^2, z \text{ central}, x^y = xz \rangle$	$\langle z \rangle$	$\langle z, x^2 \rangle$

[Recall that the notation x^y is used for the conjugate yxy^{-1}.]

7. There are two abelian groups of order 18, namely, C_{18} and $C_6 \times C_3$. For the non-abelian groups of order 18, the Sylow 2-subgroup is cyclic.

There are three non-abelian groups, $D(9)$, the direct product $D(3) \times C_3$ and the group E with presentation

$$\langle x, y, z : x^3 = y^3 = z^2 = 1, xy = yx, zxz^{-1} = x^2, zyz^{-1} = y^2 \rangle.$$

This is a pullback $D(3) \times^{\vartheta} D(3)$, using the identity automorphism on the quotient $D(3)/D(3)'$. These groups are summarised in Table B.3 below.

Table B.3 Non-abelian groups of order 18

Group	G'	$Z(G)$	Sylow 3-subgroup
$D(9)$	C_9	$\{1\}$	C_9
$D(3) \times C_3$	C_3	C_3	$C_3 \times C_3$
$D(3) \times^{\vartheta} D(3)$	$C_3 \times C_3$	$\{1\}$	$C_3 \times C_3$

8. The abelian groups of order 20 are C_{20} and $C_{10} \times C_2$. For the non-abelian groups of order 20, the Sylow 5-subgroup is cyclic. The essential properties of these groups may be seen below.

Table B.4 Non-abelian groups of order 20

Group	G'	$Z(G)$	Sylow 2-subgroup
$D(10)$	C_5	C_2	$C_2 \times C_2$
Q_{20}	C_5	C_2	C_4
$\langle x, y : x^5 = 1 = y^4,\ x^y = x^2 \rangle$	C_5	$\{1\}$	C_4

9. The only abelian group of order 21 is cyclic and the non-abelian group of order 21 has the following presentation:

$$\langle x, y : x^7 = 1 = y^3, x^y = x^2 \rangle.$$

The commutator subgroup is $\langle x \rangle$ and the group has trivial centre.

10. The abelian groups of order 24 are C_{24}, $C_{12} \times C_2$ and $C_6 \times C_2 \times C_2$. For the non-abelian groups of order 24, the Sylow 3-subgroup is cyclic. These groups are shown in Table B.5 below.

Table B.5 Non-abelian groups of order 24

Group	G'	$Z(G)$	Sylow 2-subgroup
$\langle x, y : x^3 = 1 = y^8, x^y = x^{-1} \rangle$	$\langle x \rangle$	$\langle y^2 \rangle$	C_8
$C_4 \times D(3)$	C_3	C_4	$C_4 \times C_2$
$C_2 \times Q_{12}$	C_3	$C_2 \times C_2$	$C_4 \times C_2$
$C_2 \times D(6)$	C_3	$C_2 \times C_2$	$(C_2)^3$
$C_2 \times A(4)$	$C_2 \times C_2$	C_2	$(C_2)^3$
$C_3 \times D(4)$	C_2	C_6	$D(4)$
$D(12)$	C_6	C_2	$D(4)$
$S(4)$	$A(4)$	$\{1\}$	$D(4)$
Q_{24}	C_6	C_2	Q_8
$SL(2,3)$	Q_8	C_2	Q_8
$C_3 \times Q_8$	C_2	C_6	Q_8
T	$\langle z, x^2 \rangle$	$\langle x^2 \rangle$	$D(4)$

Here T is the group of order 24 generated by x, y, z with

$$x^4 = y^2 = 1,\ yxy^{-1} = x^{-1}$$

so that $\langle x, y \rangle$ is isomorphic to the dihedral group $D(4)$, and z is an element of order 3 with $z^x = z^{-1}$ and $z^y = z$.

11. The abelian groups of order 27 are C_{27}, $C_9 \times C_3$ and $C_3 \times C_3 \times C_3$. The non-abelian groups of order 27 have presentations as follows:

$$\langle x, y : x^9 = 1 = y^3,\ x^y = x^4 \rangle$$

and

$$\langle x, y, z : x^3 = 1 = y^3 = z^3,\ z \text{ central},\ x^y = xz \rangle.$$

In both cases $G' = Z(G)$ is cyclic of order 3.

12. The abelian groups of order 28 are C_{28} and $C_{14} \times C_2$. For the non-abelian groups of order 28, the Sylow 7-subgroup is cyclic. The non-abelian groups are $D(14)$ and Q_{28} (see Table B.6 below).

Table B.6 Non-abelian groups of order 28

Group	G'	$Z(G)$	Sylow 2-subgroup
$D(14)$	C_7	C_2	$C_2 \times C_2$
Q_{28}	C_7	C_2	C_4

13. The only abelian group of order 30 is cyclic. In the non-abelian groups of order 30, all the Sylow subgroups are cyclic. The groups are listed below.

Table B.7 Non-abelian groups of order 30

Group	G'	$Z(G)$
$D(15)$	C_{15}	$\{1\}$
$D(5) \times C_3$	C_5	C_3
$D(3) \times C_5$	C_3	C_5

C

Solutions to exercises

Chapter 1

1. (a) Subtraction is not an associative operation: for example

$$(1-2)-3=-4, \text{ but } 1-(2-3)=2.$$

It follows that this set is not a group.

(b) The set is a group under this operation. Clearly (G1) is satisfied. For (G2),

$$\begin{aligned}(a \circ b) \circ c = (a+b+2) \circ c &= a+b+c+4 \\ &= a \circ (b+c+2) = a \circ (b \circ c).\end{aligned}$$

The identity is -2 since

$$a \circ (-2) = a + (-2) + 2 = a = (-2) \circ a, \text{ for all } a.$$

Finally, the inverse of a is $-(a+4)$ since

$$a \circ (-(a+4)) = a - (a+4) + 2 = -2 = -(a+4) \circ a.$$

(c) In this example, axiom (G4) fails. For example, the multiplicative inverse of 3 is not an integer.

(d) This set is a group. The determinant of a product of matrices AB is the product of the determinant of A with that of B, so the set is closed under matrix multiplication. The other axioms follow since matrix multiplication is associative, the identity matrix has determinant 1, and the inverse of a matrix A of determinant ± 1 also has determinant ± 1.

2. The required table is

	I	A	B	C	D	E	F	G
I	I	A	B	C	D	E	F	G
A	A	B	C	I	E	F	G	D
B	B	C	I	A	F	G	D	E
C	C	I	A	B	G	D	E	F
D	D	G	F	E	I	C	B	A
E	E	D	G	F	A	I	C	B
F	F	E	D	G	B	A	I	C
G	G	F	E	D	C	B	A	I

It follows as in Example 1.8 that this set is a group, but since $AD = E$ and $DA = G$, the group is non-abelian.

3. Use multiplicative notation so that the identity element is denoted by 1. Denote rotation (anticlockwise) through 90 degrees by r, rotation through 180 degrees by s, and rotation through 270 degrees by t. Let a be reflection through the vertical axis of symmetry of the square, b denote the composite of rotation r with reflection a (so that b is reflection in the line $x = -y$), c denote the composite of s with a (so that c is reflection in the horizontal axis of symmetry) and d denote the composite of t with a (so that d is reflection in the line $x = y$). The resulting table is

	1	r	s	t	a	b	c	d
1	1	r	s	t	a	b	c	d
r	r	s	t	1	b	c	d	a
s	s	t	1	r	c	d	a	b
t	t	1	r	s	d	a	b	c
a	a	d	c	b	1	t	s	r
b	b	a	d	c	r	1	t	s
c	c	b	a	d	s	r	1	t
d	d	c	b	a	t	s	r	1

4. (a) A non-square rectangle has four symmetries, the identity 1, rotation through 180 degrees a, reflection in a vertical axis b, and reflection in a horizontal axis $c (= ab)$. The table is

	1	a	b	c
1	1	a	b	c
a	a	1	c	b
b	b	c	1	a
c	c	b	a	1

(b) A parallelogram with unequal sides has no axis of reflectional symmetry, so the only non-identity symmetry is rotation through 180 degrees. Denoting this by a, the table is

	1	a
1	1	a
a	a	1

(c) A rhombus has four symmetries. If we denote these by 1 (the identity symmetry), a as rotation through 180 degrees, b as reflection in a diagonal, and c as reflection in the other diagonal, it may be seen that the table for this group is identical with that in (a) above.

5. The two required tables are

	$(1,1)$	$(-1,1)$	$(1,\omega)$	$(-1,\omega)$	$(1,\tau)$	$(-1,\tau)$
$(1,1)$	$(1,1)$	$(-1,1)$	$(1,\omega)$	$(-1,\omega)$	$(1,\tau)$	$(-1,\tau)$
$(-1,1)$	$(-1,1)$	$(1,1)$	$(-1,\omega)$	$(1,\omega)$	$(-1,\tau)$	$(1,\tau)$
$(1,\omega)$	$(1,\omega)$	$(-1,\omega)$	$(1,\tau)$	$(-1,\tau)$	$(1,1)$	$(-1,1)$
$(-1,\omega)$	$(-1,\omega)$	$(1,\omega)$	$(-1,\tau)$	$(1,\tau)$	$(-1,1)$	$(1,1)$
$(1,\tau)$	$(1,\tau)$	$(-1,\tau)$	$(1,1)$	$(-1,1)$	$(1,\omega)$	$(-1,\omega)$
$(-1,\tau)$	$(-1,\tau)$	$(1,\tau)$	$(-1,1)$	$(1,1)$	$(-1,\omega)$	$(1,\omega)$

and

	$(1,1)$	$(\omega,1)$	$(\tau,1)$	$(1,\omega)$	(ω,ω)	(τ,ω)	$(1,\tau)$	(ω,τ)	(τ,τ)
$(1,1)$	$(1,1)$	$(\omega,1)$	$(\tau,1)$	$(1,\omega)$	(ω,ω)	(τ,ω)	$(1,\tau)$	(ω,τ)	(τ,τ)
$(\omega,1)$	$(\omega,1)$	$(\tau,1)$	$(1,1)$	(ω,ω)	(τ,ω)	$(1,\omega)$	(ω,τ)	(τ,τ)	$(1,\tau)$
$(\tau,1)$	$(\tau,1)$	$(1,1)$	$(\omega,1)$	(τ,ω)	$(1,\omega)$	(ω,ω)	(τ,τ)	$(1,\tau)$	(ω,τ)
$(1,\omega)$	$(1,\omega)$	(ω,ω)	(τ,ω)	$(1,\tau)$	(ω,τ)	(τ,τ)	$(1,1)$	$(\omega,1)$	$(\tau,1)$
(ω,ω)	(ω,ω)	(τ,ω)	$(1,\omega)$	(ω,τ)	(τ,τ)	$(1,\tau)$	$(\omega,1)$	$(\tau,1)$	$(1,1)$
(τ,ω)	(τ,ω)	$(1,\omega)$	(ω,ω)	(τ,τ)	$(1,\tau)$	(ω,τ)	$(\tau,1)$	$(1,1)$	$(\omega,1)$
$(1,\tau)$	$(1,\tau)$	(ω,τ)	(τ,τ)	$(1,1)$	$(\omega,1)$	$(\tau,1)$	$(1,\omega)$	(ω,ω)	(τ,ω)
(ω,τ)	(ω,τ)	(τ,τ)	$(1,\tau)$	$(\omega,1)$	$(\tau,1)$	$(1,1)$	(ω,ω)	(τ,ω)	$(1,\omega)$
(τ,τ)	(τ,τ)	$(1,\tau)$	(ω,τ)	$(\tau,1)$	$(1,1)$	$(\omega,1)$	(τ,ω)	$(1,\omega)$	(ω,ω)

where τ denotes ω^2 and $\omega^3 = 1$.

6. If G and H are both abelian then

$$(g_1, h_1)(g_2, h_2) = (g_1 g_2, h_1 h_2) = (g_2 g_1, h_2 h_1) = (g_2, h_2)(g_1, h_1),$$

so that $G \times H$ is abelian. Conversely, if $G \times H$ is abelian, then for all g_1 and g_2 in G,

$$(g_1 g_2, 1) = (g_1, 1)(g_2, 1) = (g_2, 1)(g_1, 1) = (g_2 g_1, 1).$$

It follows that $g_1 g_2 = g_2 g_1$, so that G is abelian. A similar proof shows that H is abelian.

Chapter 2

1. There are eight maps from X to Y:

$$f_1(a) = u, f_1(b) = u, f_1(c) = u; \;\; f_2(a) = u, f_2(b) = u, f_2(c) = v;$$
$$f_3(a) = u, f_3(b) = v, f_3(c) = u; \;\; f_4(a) = u, f_4(b) = v, f_4(c) = v;$$
$$f_5(a) = v, f_5(b) = u, f_5(c) = u; \;\; f_6(a) = v, f_6(b) = u, f_6(c) = v;$$
$$f_7(a) = v, f_7(b) = v, f_7(c) = u; \;\; f_8(a) = v, f_8(b) = v, f_8(c) = v.$$

There are nine maps from Y to X:

$$g_1(u) = a, g_1(v) = a; \;\; g_2(u) = a, g_2(v) = b; \;\; g_3(u) = a, g_3(v) = c;$$
$$g_4(u) = b, g_4(v) = a; \;\; g_5(u) = b, g_5(v) = b; \;\; g_6(u) = b, g_6(v) = c;$$
$$g_7(u) = c, g_7(v) = a; \;\; g_8(u) = c, g_8(v) = b; \;\; g_9(u) = c, g_9(v) = c.$$

2. (a) Suppose that f and g are both injective and that $fg(x_1) = fg(x_2)$, so that $f(g(x_1)) = f(g(x_2))$. Since f is injective, it follows that $g(x_1) = g(x_2)$, and the fact that g is injective then implies that $x_1 = x_2$. Thus fg is injective.

(b) Suppose that f and g are both surjective. Thus for any z in Z, there is an element y in Y such that $f(y) = z$. There is then an element x in X such that $g(x) = y$, and fg is surjective since

$$fg(x) = f(g(x)) = f(y) = z,$$

To provide the required examples, first let f be the identity map on a set Y and $g : X \to Y$ be a map which is surjective but not injective, then the composite fg will not be injective. Conversely, if g is the identitiy map on X and $f : X \to Y$ is injective but not surjective, then the composite fg will not be surjective.

3. There are three maps of the required form:

$$f(a) = f(b) = f(c) = a;$$
$$g(a) = g(b) = g(c) = b;$$
$$h(a) = h(b) = h(c) = c.$$

The composition table for these maps is as follows:

	f	g	h
f	f	f	f
g	g	g	g
h	h	h	h

This is not the table of a group since there is no identity element.

4. The relation is reflexive: xRx since $2x$ is always even; the relation is symmetric since $x + y = y + x$; and the relation is transitive since if $x + y$ and $y + z$ are both even integers, then their sum $x + 2y + z$ is also even and so $x + z$ must be even. Every even integer is in the equivalence class of the integer 0, and every odd integer is in the equivalence class of the integer 1. These two classes are different, so there are just two equivalence classes.

The relation xRy if $x + y$ is divisible by 3 is not an equivalence relation since it is not reflexive: 1 is not related to itself.

5. The required tables are

$+$	$[0]_4$	$[1]_4$	$[2]_4$	$[3]_4$
$[0]_4$	$[0]_4$	$[1]_4$	$[2]_4$	$[3]_4$
$[1]_4$	$[1]_4$	$[2]_4$	$[3]_4$	$[0]_4$
$[2]_4$	$[2]_4$	$[3]_4$	$[0]_4$	$[1]_4$
$[3]_4$	$[3]_4$	$[0]_4$	$[1]_4$	$[2]_4$

and

$\times$	$[1]_5$	$[2]_5$	$[3]_5$	$[4]_5$
$[1]_5$	$[1]_5$	$[2]_5$	$[3]_5$	$[4]_5$
$[2]_5$	$[2]_5$	$[4]_5$	$[1]_5$	$[3]_5$
$[3]_5$	$[3]_5$	$[1]_5$	$[4]_5$	$[2]_5$
$[4]_5$	$[4]_5$	$[3]_5$	$[2]_5$	$[1]_5$

6. Suppose that $[a]_n=[b]_n$ and $[c]_n = [d]_n$, so that n divides $b - a$ and also $d - c$. Then

$$bd - ac = b(d - c) + c(b - a),$$

so that n divides $bd - ac$. Thus $[ac]_n = [bd]_n$, as required.

Chapter 3

1. If g is an element of any group with $g^2 = 1$, then $gg = 1$, and so g is its own inverse. If now G is a group in which $g^2 = 1$ for all g in G, then for any x and y in G, $x^{-1} = x, y^{-1} = y$ and $(xy)^{-1} = xy$. Thus

$$xy = (xy)^{-1} = y^{-1}x^{-1} = yx$$

and hence G is abelian.

2. (a) If $axa^{-1} = 1$ then $ax = a$ and so $x = a^{-1}a = 1$.

(b) If $axa^{-1} = a$ then $ax = a^2$ and so $x = a^{-1}a^2 = a$.
(c) If $axb = c$ then $ax = cb^{-1}$ and so $x = a^{-1}cb^{-1}$.
(d) If $ba^{-1}xab^{-1} = ba$ then $ba^{-1}x = baba^{-1}$ and so $x = a^2ba^{-1}$.

3. The set is closed since $xc^{-1}y$ is an element of G. To check associativity:

$$\begin{aligned} x * (y * z) = x * (yc^{-1}z) &= xc^{-1}(yc^{-1}z) \\ &= (xc^{-1}y)c^{-1}z = (xc^{-1}y) * z = (x * y) * z. \end{aligned}$$

The identity element is c since

$$c * x = cc^{-1}x = x = xc^{-1}c = x * c,$$

and the inverse of g is $cg^{-1}c$ since

$$g * (cg^{-1}c) = gc^{-1}cg^{-1}c = c = cg^{-1}cc^{-1}g = (cg^{-1}c) * g.$$

4. The order of e is 1, each of a and b have order three and the other three elements c, d and f have order 2.

5. An example can be found in the group $D(3)$. In this group, c and d each have order 2, and so are their own inverses. However $cd = a$. Thus $(cd)^{-1} = a^{-1} = b$ is different from $c^{-1}d^{-1} = cd$.

6. For any x and y in G, we have

$$(xx)(yy) = x^2y^2 = (xy)^2 = (xy)(xy).$$

Thus $xxyy = xyxy$ and so $xy = yx$, showing that G is abelian.

7. When $k = 1, x^{-1}g^1x = (x^{-1}gx)^1$. Thus suppose that

$$(x^{-1}gx)^k = x^{-1}g^kx.$$

Then

$$\begin{aligned} (x^{-1}gx)^{k+1} &= (x^{-1}gx)^k(x^{-1}gx) \quad \text{by definition} \\ &= (x^{-1}g^kx)(x^{-1}gx) \quad \text{by induction} \\ &= x^{-1}g^k(xx^{-1})gx \quad \text{using associativity} \\ &= x^{-1}g^{k+1}x, \quad \text{as required.} \end{aligned}$$

Thus if $(x^{-1}gx)^k = 1$, then $x^{-1}g^kx = 1$, so $g^kx = x1 = x$. It follows that $g^k = xx^{-1} = 1$. This shows that the order of g divides the order of $x^{-1}gx$, using Proposition 3.10.

Conversely, if $g^k = 1$, then

$$(x^{-1}gx)^k = x^{-1}g^k x = x^{-1}1x = x^{-1}x = 1.$$

Since this implies that the order of $x^{-1}gx$ divides the order of g, it follows that g and $x^{-1}gx$ have the same order.

8. It may be easily checked that

$$X^k = \begin{pmatrix} \omega^k & 0 \\ 0 & \omega^{-k} \end{pmatrix}.$$

It follows from this that $X^3 = -I$ so that $X^6 = I$ and also that

$$X^{-1} = \begin{pmatrix} \omega^{-1} & 0 \\ 0 & \omega \end{pmatrix}.$$

Thus, if $Y = \begin{pmatrix} a & b \\ c & d \end{pmatrix}$ then

$$XY = \begin{pmatrix} \omega a & \omega b \\ \omega^{-1}c & \omega^{-1}d \end{pmatrix}, \quad YX^{-1} = \begin{pmatrix} a\omega^{-1} & b\omega \\ c\omega^{-1} & d\omega \end{pmatrix},$$

and

$$Y^2 = \begin{pmatrix} a^2 + bc & b(a+d) \\ c(a+d) & d^2 + bc \end{pmatrix}.$$

So, if $XY = YX^{-1}$ and $Y^2 = X^3$, we see that

$$\omega a = \omega^{-1}a, \omega d = \omega^{-1}d, b(a+d) = 0 = c(a+d), a^2 + bc = -1 = d^2 + bc.$$

Thus, $a = d = 0$ and $bc = -1$. A suitable matrix Y is therefore

$$Y = \begin{pmatrix} 0 & b \\ -1/b & 0 \end{pmatrix} \text{ for } b \neq 0.$$

We can now form the table for G :

	I	X	X^2	X^3	X^4	X^5	Y	YX	YX^2	YX^3	YX^4	YX^5
I	I	X	X^2	X^3	X^4	X^5	Y	YX	YX^2	YX^3	YX^4	YX^5
X	X	X^2	X^3	X^4	X^5	I	YX^5	Y	YX	YX^2	YX^3	YX^4
X^2	X^2	X^3	X^4	X^5	I	X	YX^4	YX^5	Y	YX	YX^2	YX^3
X^3	X^3	X^4	X^5	I	X	X^2	YX^3	YX^4	YX^5	Y	YX	YX^2
X^4	X^4	X^5	I	X	X^2	X^3	YX^2	YX^3	YX^4	YX^5	Y	YX
X^5	X^5	I	X	X^2	X^3	X^4	YX	YX^2	YX^3	YX^4	YX^5	Y
Y	Y	YX	YX^2	YX^3	YX^4	YX^5	X^3	X^4	X^5	I	X	X^2
YX	YX	YX^2	YX^3	YX^4	YX^5	Y	X^2	X^3	X^4	X^5	I	X
YX^2	YX^2	YX^3	YX^4	YX^5	Y	YX	X	X^2	X^3	X^4	X^5	I
YX^3	YX^3	YX^4	YX^5	Y	YX	YX^2	I	X	X^2	X^3	X^4	X^5
YX^4	YX^4	YX^5	Y	YX	YX^2	YX^3	X^5	I	X	X^2	X^3	X^4
YX^5	YX^5	Y	YX	YX^2	YX^3	YX^4	X^4	X^5	I	X	X^2	X^3

The table may be obtained directly from the defining relations without using the explicit value for the matrix Y. We see from this table that I has order one, X^3 has order two, YX^i has order four (for $0 \le i \le 5$), X^2, X^4 have order three and that X, X^5 have order six.

Chapter 4

1. (a) The identity of this additive group is 0, which is even; the negative of an even integer is even and the sum of even integers is also even, so H is a subgroup.

(b) The set H is not a subgroup because it is not closed, for example $(1\ 2)(1\ 3)$ is not an element of H.

(c) The set contains the identity I (take $a = 0$); the product of two matrices of the given form is of the given form; and

$$\begin{pmatrix} 1 & a \\ 0 & 1 \end{pmatrix}^{-1} = \begin{pmatrix} 1 & -a \\ 0 & 1 \end{pmatrix},$$

so H is a subgroup.

2. In the dihedral group $D(3)$, the set $\{1, b\}$ is a subgroup, as is $\{1, ba\}$, but the set $\{1, b, ba\}$ is not closed (here the elements a and b are generators of the dihedral group as defined in the presentation on p. 33).

3. The group $\langle A, D\rangle$ contains I, A, $A^2\,(= B)$, $A^3 = C$, D, $AD = E$, $A^2D = BD = F$ and $A^3D = CD = G$, and so it is the whole group. The group $\{I, A, B, C\}$ consists of the powers of the element A of order 4, so is cyclic. Since it contains A and C, it is the smallest subgroup containing these and so $\langle A, C\rangle$ is cyclic. The table for $\langle B, F\rangle$ is

	I	B	F	D
I	I	B	F	D
B	B	I	D	F
F	F	D	I	B
D	D	F	B	I

4. We first find the number of elements in G. The group will necessarily contain the powers of x, these being $1, x, x^2, x^3$ as well as their products by y: y, yx, yx^2, yx^3. Other products, such as xy, will also need to be in G, but the relation $xy = yx^{-1}$ ensures this. In fact, consider the following list of 8 elements:

$$1,\ x,\ x^2,\ x^3,\ y,\ yx,\ yx^2,\ yx^3.$$

These clearly must be in G. To show that this list is the complete list of elements of G, we only need to check that these are a closed set. The method used to do this is similar to that used in Example 4.10 since $yx = xy^{-1}$. This gives the following table:

	1	y	y^2	y^3	x	xy	xy^2	xy^3
1	1	y	y^2	y^3	x	xy	xy^2	xy^3
y	y	y^2	y^3	1	xy^3	x	xy	xy^2
y^2	y^2	y^3	1	y	xy^2	xy^3	x	xy
y^3	y^3	1	y	y^2	xy	xy^2	xy^3	x
x	x	xy	xy^2	xy^3	y^2	y^3	1	y
xy	xy	xy^2	xy^3	x	y	y^2	y^3	1
xy^2	xy^2	xy^3	x	xy	1	y	y^2	y^3
xy^3	xy^3	x	xy	xy^2	y^3	1	y	y^2

Chapter 5

1. (a) The left cosets of $H = \langle A \rangle$ in G are:

$$H = 1H = AH = BH = CH = \{I, A, B, C\};\ \text{and}$$

$$DH = EH = FH = GH = \{D, E, F, G\}.$$

(b) The left cosets of $H = \langle B, F \rangle$ in G are:

$$H = 1H = BH = DH = FH = \{I, B, D, F\};\ \text{and}$$

$$AH = CH = EH = GH = \{A, C, E, G\}.$$

(c) The left cosets of $H = \{I, D\}$ in G are:

$$H = 1H = DH = \{I, D\};\quad AH = EH = \{A, E\};$$

$$BH = FH = \{B, F\}; \text{ and } CH = GH = \{C, G\}.$$

The right cosets of H in G are:

$$H = H1 = HD = \{I, D\}; \quad HA = HG = \{A, G\};$$

$$HB = HF = \{B, F\}; \text{ and } HC = HE = \{C, E\}.$$

2. If gH were a subgroup, it would contain the identity element 1, so there would be an element $h \in H$ such that $gh = 1$. Then g would be the inverse of h, and since H is a subgroup, we would deduce that g is in H.

3. If y is in Hx, then $y = hx$ for some $h \in H$. Then

$$Hy = Hhx = Hx.$$

4. Suppose that $Hx = Hy$, so that $hx = y$ for some $h \in H$. Thus $yx^{-1} = h$, so that yx^{-1} is an element of H. Conversely, if $yx^{-1} = h$ for some $h \in H$, then $y = hx$ and y is in the coset Hx. It then follows by Question 3 that $Hy = Hx$.

5. An example is provided by the same group as in the solution to Exercise 2 of Chapter 4. We therefore take G to be the dihedral group $D(3)$, A to be $\{1, a\}$ and B to be $\{1, b\}$, so that AB is the set $\{1, a, b, ab\}$.

6. The intersection $P \cap Q$ is a subgroup of P and so has order p^c for some c with $0 \le c \le a$. By Proposition 5.18, $|PQ| = p^{a+b-c}$. If PQ were a subgroup of G, then by Lagrange's Theorem, the order of PQ would divide $|G|$. Since the order of PQ is a power of p, and the highest power of p dividing G is p^a, it follows that $|PQ|$ could have order at most p^a. This would imply that $a+b-c \le a$, so that $b \le c$. However, $P \cap Q$ is a subgroup of Q, so this can only occur if $b = c$. It follows that $P \cap Q = Q$, showing that Q is a subgroup of P.

Chapter 6

1. The 8 codewords are

000111; 001110; 010101; 011100;
100011; 101010; 110001; 111000.

This is not a linear code because the sum of the first two codewords is 001001, which is not a codeword. The least weight of C is 3, but the minimum distance of C is 2 (between first and third words, for example).

2. (a) The eight codewords are

$$00000;\ 10001;\ 01010;\ 11011;$$
$$00111;\ 10110;\ 01101;\ 11100.$$

It follows that the minimum distance for this code is 2, so that it detects an error, but does not correct errors.

(b) There are nine codewords

$$0000;\ 1011;\ 2022;\ 0112;\ 1120;\ 2101;\ 0221;\ 1202;\ 2210.$$

Since the minimum distance for this code is 3, the code detects two errors and corrects one error.

(c) This code has 125 codewords. The sum of twice the first row of the generator matrix and its second row is 21000, so the least weight is at most 2. Consideration of the first three columns shows that there can be no codeword of weight 1, so the code detects one error, but does not correct any errors.

3. To show that C^+ is a subgroup, first note that the zero vector is in C^+ since this has weight 0. Suppose that c and d are codewords of even weights, and that c and d agree in k coordinates. The vector $c + d$ has weight

$$(wt(c) - k) + (wt(d) - k),$$

and this is an even integer. Since the negative of a vector $\mathbf{v}$ over $GF(2)$ is equal to $\mathbf{v}$, C^+ is a subgroup.

If $C \neq C^+$, there is an element c of odd weight in C. In this case, denote the elements of odd weight by C^-. Adding c to each element of C^+ in turn gives distinct codewords of odd weight, so $|C^-| \geq |C^+|$. However, adding c to each element of C^- in turn gives distinct codewords of even weight, so $|C^-| \leq |C^+|$. Since $C = C^+ \cup C^-$, it follows that $|C| = 22|C^+|$, as required.

4. The required table is

0000	1011	2022	0112	1120	2101	0221	1202	2210
0001	1012	2020	0110	1121	2102	0222	1200	2211
0002	1010	2021	0111	1122	2100	0220	1201	2212
0010	1021	2002	0122	1100	2111	0201	1212	2220
0020	1001	2012	0102	1110	2121	0211	1222	2200
0100	1111	2122	0212	1220	2201	0021	1002	2010
0200	1211	2222	0012	1020	2001	0121	1102	2110
1000	2011	0022	1112	2120	0101	1221	2202	0210
2000	0011	1022	2112	0120	1101	2221	0202	1210

5. The parity check matrix is

$$P = \begin{pmatrix} 1 & 1 & 1 & 1 & 0 & 0 & 0 \\ 1 & 1 & 0 & 0 & 1 & 0 & 0 \\ 0 & 1 & 1 & 0 & 0 & 1 & 0 \\ 1 & 0 & 1 & 0 & 0 & 0 & 1 \end{pmatrix}.$$

An example of a two column decoding table is as follows.

Syndrome	Coset leaders	Syndrome	Coset leaders
0000	0000000	0011	0000011
1101	1000000	0101	0000101
1110	0100000	1001	0001001
1011	0010000	0110	0000110
1000	0001000	1010	0001010
0100	0000100	1100	0001100
0010	0000010	1111	1000010
0001	0000001	0111	0000111

The syndrome of 1100011 is 0000 and so this is a codeword;
the syndrome of 1011000 is 1110 and so this word corrects to 1111000;
the syndrome of 0101110 is 0000 and so this is a codeword;
the syndrome of 0110001 is 0100 and so this word corrects to 0110101;
the syndrome of 1010110 is 0000 and so this is a codeword.

Chapter 7

1. Since H is a subgroup, $1 \in H$, and so $g1g^{-1} = 1 \in gHg^{-1}$.
If ghg^{-1} and gkg^{-1} are in gHg^{-1}, then

$$(ghg^{-1})(gkg^{-1}) = g(hk)g^{-1} \in gHg^{-1},$$

since $hk \in H$. Finally,

$$(ghg^{-1})^{-1} = gh^{-1}g^{-1} \in gHg^{-1}$$

since $h^{-1} \in H$.

2. The group $G = D(3)$ has 6 elements, so apart from G itself, and the identity subgroup $\{1\}$, Lagrange's Theorem tells us that each other subgroup has either 2 or 3 elements. Groups of order 2 are cyclic, as are groups of order 3, so the complete list of subgroups of

$$G = \langle x, y : x^2 = 1 = y^3 \text{ and } yx = xy^{-1} \rangle$$

is:

$$G, \langle y\rangle = \{1, y, y^2\} = \langle y^2\rangle, \quad \langle x\rangle = \{1, x\},$$

$$\langle xy\rangle = \{1, xy\}, \langle xy^2\rangle = \{1, xy^2\} \text{ and } \{1\}.$$

Of these the group $G, \langle y\rangle$ (of index two) and $\{1\}$ are normal, but the other three subgroups are not normal:

$$y\langle x\rangle \neq \langle x\rangle y;\ y\langle xy\rangle \neq \langle xy\rangle y \quad \text{and} \quad x\langle xy^2\rangle \neq \langle xy^2\rangle x.$$

3. Note that $N = \{1, x^2\}$ is a subgroup since x^2 has order 2.
The cosets are:

$$E = N = 1N = N1; \qquad A = \{x, x^3\} = xN = Nx;$$

$$B = \{y, yx^2\} = yN = Ny; \qquad C = \{yx, yx^3\} = yxN = Nyx.$$

This shows that N is a normal subgroup. Then G/N has the following table:

	E	A	B	C
E	E	A	B	C
A	A	E	C	B
B	B	C	E	A
C	C	B	A	E

4. In this case, the left coset $aH = \{a, ab\} = \{a, ba^{-1}\}$, whereas the right coset $Ha = \{a, ba\}$, so H is not normal. To see that multiplication of cosets is not well-defined, note that $aH = ba^{-1}H$, and $baH = a^{-1}H$, but $(abaH) = (ba^{-1})aH = bH = H$ and

$$(ba^{-1}H)(a^{-1}H) = (ba^{-1}a^{-1})H = ba^{-2}H = \{ba^2, a^2\}.$$

5. To show that $Z(G)$ is a subgroup of G, note that 1 is in $Z(G)$ since $1g = g1$ for all g in G. Now let w and z be in $Z(G)$. Thus for all g in $G, zg = gz$ and $wg = gw$. Then zw is in $Z(G)$ since

$$(zw)g = z(wg) = z(gw) = (zg)w = (gz)w = g(zw).$$

Also z^{-1} is in $Z(G)$ since the fact that $zg = gz$ for all g in G gives that $gz^{-1} = z^{-1}g$ for all g in G. Thus $Z(G)$ is a subgroup of G. Also $Z(G)$ is abelian since if z and w are in $Z(G)$, then since w is in $G, zw = wz$. To show that $Z(G)$ is normal, take $g \in G$ and $z \in Z(G)$, then

$$g^{-1}zg = g^{-1}gz = z \in Z(G).$$

When G is $D(3)$, neither x nor y is in $Z(G)$ since $xy \neq yx$, also y^2 and xy are not in $Z(G)$ since $y^2xy = xy^2 \neq xyy^2$, and xy^2 is not in $Z(G)$ since $xxy^2 \neq xy^2x$, so $Z(G) = \{1\}$. A similar calculation in $D(4)$ shows that the only candidates for $Z(D(4))$ are 1 and y^2. It can be checked that y^2 does indeed commute with each of the eight elements in G, so $Z(G) = \{1, y^2\}$.

Chapter 8

1. Suppose that ϕ is injective. By Corollary 8.7, 1_G is in $\ker\phi$, and since ϕ is injective, no other element of G satisfies $\phi(g) = 1_H$, so $\ker\phi = \{1\}$. Conversely, suppose that $\ker\phi = \{1\}$ and $\phi(x) = \phi(y)$. Then

$$1_H = \phi(y)^{-1}\phi(x) = \phi(y^{-1}x),$$

using Proposition 8.6. Thus $y^{-1}x \in \ker\phi$, so $y^{-1}x = 1$ and $x = y$, showing that ϕ is injective.

2. The fact that ϑ is a homomorphism follows because the composite of two reflections (or of two rotations) is a rotation, whereas the composite of a reflection and a rotation is a reflection. The kernel of ϑ is the subgroup of the three rotations, and the image of ϑ is the set $\{1, -1\}$.

3. To show that ϕ is a homomorphism, consider

$$\begin{aligned}
\phi(g_1, g_2)\phi(g_3, g_4) &= \vartheta(g_1)\vartheta(g_2)^{-1}\vartheta(g_3)\vartheta(g_4)^{-1} \\
&= \vartheta(g_1)\vartheta(g_3)\vartheta(g_2)^{-1}\vartheta(g_4)^{-1} \quad \text{since } H \text{ is abelian} \\
&= \vartheta(g_1g_3)\vartheta(g_2^{-1}g_4^{-1}) \\
&= \vartheta(g_1g_3)\vartheta((g_2g_4)^{-1}) \quad \text{since } H \text{ is abelian} \\
&= \phi(g_1g_3, g_2g_4),
\end{aligned}$$

so that ϕ is a homomorphism. When $G = D(3)$, there are 36 elements in $G \times G$. Of these the following 18 pairs $(g_1\ g_2)$ with g_1 and g_2 either both rotations or both reflections are in $\ker\phi$:

$$(1,1), (1,a), (1,a^2), (a,1), (a,a), (a,a^2),$$

$$(a^2,1), (a^2,a), (a^2,a^2), (b,b), (b,ba), (b,ba^2),$$

$$(ba, b), (ba, ba), (ba, ba^2), (ba^2, b), (ba^2, ba), (ba^2, ba^2),$$

where $1, a$ and a^2 denote the rotations in $D(3)$ and b, ba, ba^2 denote the reflections.

4. When $k = 1, \phi(g) = \phi(g)$. Suppose that $\phi(g^k) = \phi(g)^k$, then

$$\phi(g^{k+1}) = \phi(g^k g) = \phi(g)^k \phi(g) = \phi(g)^{k+1}.$$

If g has order k, we see that $1 = \phi(1) = \phi(g^k) = \phi(g)^k$, so that if r is the order of $\phi(g)$ then r divides k, by Proposition 3.10. Since

$$\phi(1) = 1 = \phi(g)^r = \phi(g^r),$$

we see that if ϕ is injective, then r divides k and so $k = r$.

5. An automorphism of C_3 is determined by its value on ω. Since this cannot be 1 if the map is to be injective, there are just two possible values, giving two maps;

$$\phi(1) = 1;\ \phi(\omega) = \omega;\ \phi(\omega^2) = \omega^2; \text{ and}$$
$$\vartheta(1) = 1;\ \vartheta(\omega) = \omega^2;\ \vartheta(\omega^2) = \omega.$$

The multiplication table for $\mathrm{Aut}(C_3)$ is the following.

	ϕ	ϑ
ϕ	ϕ	ϑ
ϑ	ϑ	ϕ

Chapter 9

1. In cycle notation

$$\pi = (1\ 3\ 5\ 7)(2\ 4\ 6) \quad \text{and} \quad \rho = (1\ 8)(2\ 7)(3\ 6)(4\ 5).$$

It follows that

$$\pi\rho = (1\ 8\ 3\ 2)(4\ 7)(5\ 6); \quad \rho\pi = (1\ 6\ 7\ 8)(2\ 5)(3\ 4); \quad \text{and}$$

$$\pi^2\rho = (1\ 8\ 5\ 2\ 3\ 4)(6\ 7).$$

Thus π has order 12 and is odd, ρ has order 2 and is even and $\pi\rho$ has order 4 and is odd.

2. The elements of $S(4)$ are

$$1;$$

$$(1\ 2), (1\ 3), (1\ 4), (2\ 3), (2\ 4), (3\ 4);$$

$$(1\ 2\ 3), (1\ 3\ 2), (1\ 2\ 4), (1\ 4\ 2), (1\ 3\ 4), (1\ 4\ 3), (2\ 3\ 4), (2\ 4\ 3);$$

$$(1\ 2\ 3\ 4), (1\ 4\ 3\ 2), (1\ 3\ 2\ 4), (1\ 4\ 2\ 3), (1\ 2\ 4\ 3), (1\ 3\ 4\ 2);$$

$$(1\ 2)(3\ 4), (1\ 3)(2\ 4), (1\ 4)(2\ 3).$$

To show that V is a subgroup consider the table:

	1	(1 2)(3 4)	(1 3)(2 4)	(1 4)(2 3)
1	1	(1 2)(3 4)	(1 3)(2 4)	(1 4)(2 3)
(1 2)(3 4)	(1 2)(3 4)	1	(1 4)(2 3)	(1 3)(2 4)
(1 3)(2 4)	(1 3)(2 4)	(1 4)(2 3)	1	(1 2)(3 4)
(1 4)(2 3)	(1 4)(2 3)	(1 3)(2 4)	(1 2)(3 4)	1

Since V is closed under multiplication, and each conjugate of an element of V is in V (since each non-identity element of V is a product of two disjoint transpositions), we see that V is a normal subgroup of $S(4)$. Now let the cosets of V in $S(4)$ be denoted as follows:

$$\begin{aligned}
E = 1V &= \{1, (1\ 2)(3\ 4), (1\ 3)(2\ 4), (1\ 4)(2\ 3)\},\\
A = (1\ 2\ 3)V &= \{(1\ 2\ 3), (1\ 3\ 4), (2\ 4\ 3), (1\ 4\ 2)\},\\
B = (1\ 3\ 2)V &= \{(1\ 3\ 2), (2\ 3\ 4), (1\ 2\ 4), (1\ 4\ 3)\},\\
C = (2\ 3)V &= \{(2\ 3), (1\ 3\ 4\ 2), (1\ 2\ 4\ 3), (1\ 4)\},\\
D = (1\ 3)V &= \{(1\ 3), (1\ 2\ 3\ 4), (2\ 4), (1\ 4\ 3\ 2)\},\\
F = (1\ 2)V &= \{(1\ 2), (3\ 4), (1\ 3\ 2\ 4), (1\ 4\ 2\ 3)\}.
\end{aligned}$$

The resulting table is

	E	A	B	C	D	F
E	E	A	B	C	D	F
A	A	B	E	F	C	D
B	B	E	A	D	F	C
C	C	D	F	E	A	B
D	D	F	C	B	E	A
F	F	C	D	A	B	E

3. In $A(4)$ (a group with 12 elements), the identity element has order 1, the elements (1 2)(3 4), (1 3)(2 4) and (1 4)(2 3) have order 2, and the eight remaining elements (1 2 3), (1 3 2), (1 2 4), (1 4 2), (1 3 4), (1 4 3),

(2 3 4) and (2 4 3) have order 3. It follows that apart from $\{1\}$ and $A(4)$, the group has the following eight subgroups:

$$\begin{array}{l}\{1,(1\ 2)(3\ 4)\};\ \ \{1,(1\ 3)(2\ 4)\};\ \ \{1,(1\ 4)(2\ 3)\};\\ \{1,(1\ 2)(3\ 4),\ (1\ 3)(2\ 4),\ (1\ 4)(2\ 3)\};\\ \{1,(1\ 2\ 3),(1\ 3\ 2)\};\ \ \{1,(1\ 2\ 4),(1\ 4\ 2)\};\\ \{1,(1\ 3\ 4),(1\ 4\ 3)\};\{1,(2\ 3\ 4),(2\ 4\ 3)\}.\end{array}$$

It is clear that $A(4)$ does not have a subgroup with six elements, since such a subgroup would be normal (being of index two) and so would need to contain all conjugates of any of its elements of order 2. Thus this subgroup would contain the subgroup V of Exercise 2, and so would have order divisible by 4. This shows that the converse of Lagrange's Theorem does not hold.

Chapter 10

1. To show that V is a G-set:

$$\begin{array}{rcl} id \cdot v & = & \lambda_1 v_{id(1)} + \cdots + \lambda_n v_{id(n)} = \lambda_1 v_1 + \cdots + \lambda_n v_n = v; \\ \rho\pi \cdot v & = & \lambda_1 v_{\rho\pi(1)} + \cdots + \lambda_n v_{\rho\pi(n)} \\ & = & \rho \cdot (\lambda_1 v_{\pi(1)} + \cdots + \lambda_n v_{\pi(n)}) = \rho \cdot (\pi \cdot v). \end{array}$$

(a) When $v = v_1 + v_2 + v_3 + v_4$, the stabiliser of v is $S(4)$ and the orbit of v is $\{v\}$.
(b) When $v = v_1 + v_3$, the orbit of v is the set

$$\{v_1 + v_3,\ v_1 + v_2,\ v_1 + v_4,\ v_2 + v_3,\ v_2 + v_4,\ v_3 + v_4\},$$

and the stabiliser is the subgroup

$$\{1,\ (1\ 3),(2\ 4),(1\ 3)(2\ 4)\}.$$

2. The stabiliser of $g \cdot x$ is the set

$$\begin{array}{rcl} \{h \in G : h \cdot (g \cdot x) = g \cdot x\} & = & \{h \in G : hg \cdot x = g \cdot x\} \\ & = & \{h \in G : g^{-1}hg \cdot x = x\} \\ & = & \{h \in G : g^{-1}hg \in G_x\} = gG_xg^{-1}. \end{array}$$

3. The conjugacy classes in $S(4)$ are

$$\{1\}, \{(1\ 2), (1\ 3), (1\ 4), (2\ 3), (2\ 4), (3\ 4)\},$$

$$\{(1\ 2\ 3), (1\ 3\ 2), (2\ 3\ 4), (2\ 4\ 3), (1\ 2\ 4), (1\ 4\ 2), (1\ 3\ 4), (1\ 4\ 3)\},$$

$$\{(1\ 2\ 3\ 4), (1\ 4\ 3\ 2), (1\ 3\ 2\ 4), (1\ 4\ 2\ 3), (1\ 2\ 4\ 3), (1\ 3\ 4\ 2)\},$$

$$\{(1\ 2)(3\ 4), (1\ 3)(2\ 4), (1\ 4)(2\ 3)\}.$$

(See the discussion at the end of Chapter 9). A typical calculation to support these claims is as follows. To determine the class of (1 2), use Proposition 9.20, to see that conjugating (1 2) by any of the permutations taking 1 to i and 2 to j gives $(i\ j)$, so every transposition is conjugate to (1 2). Also, conjugate elements have the same cycle structure, so (1 2) can have no other conjugates.

The conjugacy classes in $A(4)$ are

$$\{1\}, \{(1\ 2\ 3), (1\ 4\ 2), (2\ 4\ 3), (1\ 3\ 4)\},$$

$$\{(1\ 3\ 2), (1\ 2\ 4), (2\ 3\ 4), (1\ 4\ 3)\},$$

$$\{(1\ 2)(3\ 4), (1\ 3)(2\ 4), (1\ 4)(2\ 3)\}.$$

The reason why not all 3-cycles are conjugate is that, by inspection, any permutation which conjugates (1 2 3) to (1 3 2) is an odd permutation and so is not in $A(4)$.

4. Since $g1 = 1g$, we see that $1 \in C_G(g)$; suppose that x, y centralise g, so that $xg = gx$ and $yg = gy$, then

$$(xy)g = x(yg) = x(gy) = (xg)y = (gx)y = g(xy)$$

so $xy \in C_G(g)$; also if $xg = gx$ then $gx^{-1} = x^{-1}g$ so $x^{-1} \in C_G(g)$.

5. In any group the identity element always forms a conjugacy class on its own, so if G has precisely two conjuagcy classses, all the non-identity elements lie in one class. Since the number of elements in a conjugacy class divides $n = |G|$, we deduce that $n - 1$ divides n. It then follows that $n - 1$ divides $n - (n - 1) = 1$ so that $n - 1 = 1$ and hence $n = 2$.

6. (a) The subgroup H is normal in G and so $N_G(H) = G$.

(b) The given subgroup is not normal in G, and since $N_G(H)$ contains H and is not G, Lagrange's Theorem implies that $N_G(H) = H$.

Chapter 11

1. (a) n_7 divides 28 and is $\equiv 1 \bmod 7$, so is 1;
   (b) n_2 divides 48 and is odd, so is 1 or 3;
   (c) G is a 2-group, so Sylow's Theorems tell us nothing;

(d) n_2 divides 12 and is odd so is 1 or 3;
(e) n_3 divides 12 and is $\equiv 1 \bmod 3$, so is 1 or 4.

2. To show that $H_1 \times H_2$ is a subgroup of $G_1 \times G_2$, note that since H_i is a subgroup of G_i $(i = 1, 2)$ then
(a) (1, 1) is in $H_1 \times H_2$;
(b) if x_i and y_i are in H_i $(i = 1, 2)$ then $(x_1y_1, x_2y_2) \in H_1 \times H_2$;
(c) $(x_1, x_2)^{-1} = (x_1^{-1}, x_2^{-1})$ and this is in $H_1 \times H_2$ since each $x_i^{-1} \in H_i$.

If H_i is a normal subgroup of G_i, then for all $g_i \in G_i$ $(i = 1, 2)$,

$$(g_1, g_2)(x_1, x_2)(g_1, g_2)^{-1} = (g_1x_1g_1^{-1}, g_2x_2g_2^{-1}) \in H_1 \times H_2.$$

If H_i is a Sylow p-subgroup of G_i $(i = 1, 2)$, then H_1 has order p^r, say, and H_2 has order p^s, say, where p^r is the highest power of p dividing $|G_1|$ and p^s is the highest power of p dividing $|G_2|$. Since $|H_1 \times H_2| = p^{r+s}$, the highest power of p dividing $|G_1 \times G_2|$, it follows that $H_1 \times H_2$ is a Sylow p-subgroup of $G_1 \times G_2$. Finally, if H_i is the unique Sylow p-subgroup of G_i $(i = 1, 2)$, then H_i is normal in G_i, and so $H_1 \times H_2$ is a normal subgroup of $G_1 \times G_2$ and hence its unique Sylow p-subgroup.

3. Let G be the group $A(5)$ with Sylow 2-subgroup consisting of the four elements $\{1, (1\ 2)(4\ 5), (1\ 4)(2\ 5), (1\ 5)(2\ 4)\}$ and let H be the subgroup $A(4)$ of even permutations on the numbers $\{1, 2, 3, 4\}$. Then $P \cap H = \{1\}$ which is not a Sylow 2-subgroup of H.

Chapter 12

1. Let G have order 35. The number of Sylow 5-subgroups is one of 1, 6, 11, 16, ... and divides 35, so is 1. There is a unique Sylow 5-subgroup P, say. The number of Sylow 7-subgroups is one of 1, 8, 15, 22, ... and divides 35, so is 1. There is a unique Sylow 7-subgroup Q, say. Then $P \cap Q$ is a subgroup of P (a group with 5 elements) and also a subgroup of Q (a group with 7 elements). It follows by Lagrange's Theorem that $P \cap Q$ can only have one element so that $P \cap Q = \{1\}$. Now let $P = \langle a \rangle$ and $Q = \langle b \rangle$ (both are cyclic by Proposition 5.19). Then $b^{-1}a^{-1}b \in P$ since P is a normal subgroup, and so $b^{-1}a^{-1}ba \in P$. Also $a^{-1}ba \in Q$ since Q is a normal subgroup, and so $b^{-1}a^{-1}ba \in Q$. Thus $b^{-1}a^{-1}ba \in P \cap Q = \{1\}$. This gives that $b^{-1}a^{-1}ba = 1$ and so $ab = ba$. It then follows by Proposition 12.2 that $(ab)^7 = a^7b^7 = a^7 \neq 1$, and $(ab)^5 = a^5b^5 = b^5 \neq 1$. By Proposition 5.12 the order of ab divides 35. Since we have shown that this order is not 1, 5 or 7, it must be 35 and so G is cyclic.

2. Let G have $105 = 3 \times 5 \times 7$ elements. The number of Sylow 7-subgroups is 1 or 15 and the number of Sylow 5-subgroups is 1 or 21. If there are 15 Sylow

7-subgroups, these contain $15 \times 6 = 90$ elements of G since the intersection of any two is $\{1\}$. In this case there could not be 21 Sylow 5-subgroups. Thus G either has 1 Sylow 7-subgroup or 1 Sylow 5-subgroup.

Suppose that G has a unique Sylow 5-subgroup P, say. Then P is a normal subgroup of G and G/P has 21 elements. By Proposition 12.1, G/P has one Sylow 7-subgroup N/P, say, so that N is a normal subgroup of G (by the Correspondence Theorem) of order $7 \times 5 = 35$. It follows by Exercise 1 that N is a normal cyclic subgroup of G of order 35. Similarly, suppose that G has a unique Sylow 7-subgroup Q, say. Then Q is a normal subgroup of G and G/Q has 15 elements. By Proposition 12.1, G/Q has one Sylow 5-subgroup M/Q, say, so that M is a normal subgroup of G (also by the Correspondence Theorem) of order $7 \times 5 = 35$. By Exercise 1, M is cyclic.

3. We may suppose that the number of Sylow 7-subgroups of our group is greater than 1, in which case it is 8. Let the Sylow 7-subgroups be $S_1, S_2, \ldots, S_8$. Hence if i is different from j, then $S_i \cap S_j$ (being a subgroup of a group with 7 elements) is $\{1\}$. Thus the 8 Sylow 7-subgroups contain $8 \times 6 = 48$ non-identity elements. Only 8 elements remain, and a Sylow 2-subgroup has 8 elements, so there can only be one Sylow 2-subgroup.

Chapter 13

1. Suppose that $G = D(4)$ were an internal direct product of two proper subgroups H and K. Without loss of generality, we may suppose that H has order 2 and K has order 4. Since H would be normal in G, the non-identity element of H would be conjugated to itself by every element of G. This means that the two elements of H would both be in the centre of G. This uniquely specifies H as $\{1, a^2\}$. Since every non-identity normal subgroup of G contains a^2, it is then impossible for $H \cap K$ to be $\{1\}$. Thus G cannot be a direct product.

2. Let z denote x^5. Then

$$yzy^{-1} = yx^5y^{-1} = (yxy^{-1})^5 = (x^4)^5 = x^{20} = x^5 = z,$$

so that z commutes with y. Since z is a power of x, z also commutes with x and so z is the centre of G. It follows that $\langle z \rangle$ is a normal subgroup of G of order 3.

Now let N be the subgroup of G generated by x^3 and y. Note that since

$$yx^3y^{-1} = (yxy^{-1})^3 = (x^4)^3 = x^{12} = x^{-3},$$

N is isomorphic to the dihedral group $D(3)$. It now follows that we only need to show that G is an internal direct product of N and $\langle z\rangle$. Since $N \cap \langle z\rangle = \{1\}$, Proposition 13.5 implies that we only need to prove that N is a normal subgroup of G to deduce that G is an internal direct product of N and $\langle z\rangle$. Since N has 10 elements, its normaliser has order divisible by 10 and dividing 30. If we show that x normalises N, it will then follow that the normaliser has order greater than 10, and so N would be a normal subgroup of G. To show that x normalises N, note that

$$xx^3x^{-1} = x^3; \text{ and } xyx^{-1} = yx^4x^{-1} = xy^3 \in N.$$

Since $G = N\langle z\rangle$, Proposition 13.5 now proves the result.

3. Given $G = D(6) = \langle a, b : a^6 = 1 = b^2, ab = ba^{-1}\rangle$, let $H = \{1, a^3\}$ and $K = \langle a^2, b\rangle$. It can then be checked, as in the solution to Exercise 2, that $H \cap K = \{1\}$, that H and K are each normal in G, that $G = HK$ and that K is isomorphic to the dihedral group $D(3)$.

4. Let G_1 and G_2 both be the dihedral group of order 8 generated by elements a_i of order 4 and b_i of order 2, so that the centre of G_i is $\{1, a_i^2\}$. In $D(4)$, a_i and a_i^3 have order 4 and all other non-identity elements have order 2. The map ϑ is the identity since $Z(G_1) = Z(G_2) = C_2$. The group $G_1 \times G_2$ has 64 elements, and the set X consists of $(1,1)$ and (a_1^2, a_2^2). Thus the central product has 32 elements.

The elements of order 4 in the central product are therefore of the form $(a_1, g_2)X$ or $(g_1, a_2)X$ as g_i varies over the elements of $D(4)$ which satisfy $g_i^2 = 1$. This follows since for such an element, the cosets $(a_1, g_2)X$ and $(a_1^3, a_2^2g_2)X$ are equal. There are therefore 12 elements of order 4, leaving 19 elements of order 2.

If G_1 and G_2 are both the quaternion group, so that the central subgroup X consists of the elements $\{(1_1, 1_2), (x_1^2, x_2^2)\}$, a similar calculation to the above shows that there are the following 19 elements of order 2 in the central product:

$$(1, x_2^2)X,$$
$$(x_1, x_2)X, (x_1, x_2^3)X, (x_1, y_2)X,$$
$$(x_1, y_2^3)X, (x_1, x_2y_2)X, (x_1, x_2^3y_2^3)X$$
$$(y_1, x_2)X, (y_1, x_2^3)X, (y_1, y_2)X,$$
$$(y_1, y_2^3)X, (y_1, x_2y_2)X, (y_1, x_2^3y_2^3)X$$
$$(x_1y_2, x_2)X, (x_1y_1, x_2^3)X, (x_1y_1, y_2)X,$$
$$(x_1y_1, y_2^3)X, (x_1y_1, x_2y_2)X, (x_1y_1, x_2^3y_2^3)X.$$

5. The 24 elements in the pullback are:

$$(1,1),\ (1,(1\ 2)(3\ 4)),\ (1,(1\ 3)(2\ 4)),\ (1,(1\ 4)(2\ 3));$$

$$(x^3,1),\ (x^3,(1\ 2)(3\ 4)),\ (x^3,(1\ 3)(2\ 4)),\ (x^3,(1\ 4)(2\ 3));$$

$$(x,(1\ 2\ 3)),\ (x,(1\ 3\ 4)),\ (x,(2\ 4\ 3)),\ (x,(1\ 4\ 2));$$

$$(x^4,(1\ 2\ 3)),\ (x^4,(1\ 3\ 4)),\ (x^4,(2\ 4\ 3)),\ (x^4,(1\ 4\ 2));$$

$$(x^2,(1\ 3\ 2)),\ (x^2,(1\ 4\ 3)),\ (x^2,(2\ 3\ 4)),\ (x^2,(1\ 2\ 4));$$

$$(x^5,(1\ 3\ 2)),\ (x^5,(1\ 4\ 3)),\ (x^5,(2\ 3\ 4)),\ (x^5,(1\ 2\ 4)).$$

Chapter 14

1. Since $360 = 2^3 3^2 5$, the list of abelian groups with 360 elements is:

$$C_{360};\ C_{180} \times C_2;\ C_{90} \times C_2 \times C_2;$$

$$C_{120} \times C_3;\ C_{60} \times C_6;\ C_{30} \times C_6 \times C_2.$$

2. Let P be a finite abelian p-group, and Q denote the set of elements of P whose order divides p^r. Since the identity element has order 1, it is in Q. If x and y are in Q then $x^{p^r} = y^{p^r} = 1$. But then, since P is abelian, by Proposition 14.2,

$$(xy)^{p^r} = x^{p^r} y^{p^r} = 1,$$

so xy is also in Q. Finally, since $(x^{-1})^{p^r} = (x^{p^r})^{-1}$, x^{-1} is also in Q, so Q is a subgroup of G.

Let P be the dihedral group of order 8. This group has two elements of order 4, 5 elements of order 2 and 1 element of order 1. It is clear from Lagrange's Theorem that the elements of order dividing 2 do not form a subgroup of this 2-group.

3. It is sufficient to prove the result for p-groups since we can then factorise d as a product of prime powers, and take the internal direct product of the various p-subgroups of appropriate sizes. By Proposition 4.14, every cyclic p-group has subgroups of every possible divisor of the group order. To generalise to any p-group, suppose that $r > n_1$, let p^r be a divisor of the order of P, and let P have type $(n_1, n_2, \ldots, n_s)$. There is an integer $k \geq 1$ such that

$$n_1 + n_2 + \cdots + n_k \leq r, \text{ but } n_1 + n_2 + \cdots + n_{k+1} > r.$$

Taking $k = 0$ when $r \leq n_1$, the cyclic subgroup of order $p^{n_{k+1}}$ has a subgroup H, say, of order p^t, where

$$t = r - (n_1 + n_2 + \cdots + n_k), \text{ with } t = r \text{ when } k = 0.$$

So $C_{p^{n_1}} \times \cdots \times C_{p^{n_k}} \times H$ is a subgroup of P of order p^r, as required.

4. There are two types of group of order p^2: cyclic groups and elementary abelian groups. Thus if x is a generator for the cyclic group of order p in $C_p \times C_{p^2}$, and y is a generator for the subgroup of order p^2, cyclic subgroups of order p^2 are generated by elements of order p^2. Since there is a unique element of the form $x^i y$ in each such subgroup, it follows that there are p cyclic subgroups of order p^2. The subgroup of elements of order dividing p is generated by x and y^p, and so has p^2 elements. This is therefore the unique elementary abelian subgroup of G. Thus there are $p^2 + 1$ subgroups with p^2 elements.

5. If G is elementary abelian of order p^n, every non-identity element generates a subgroup of order p containing $p - 1$ non-identity elements. Since any two of these subgroups are either equal or disjoint, the number of such subgroups is

$$(p^n - 1)/(p - 1) = p^{n-1} + \cdots + p + 1.$$

Any group of the form $C_{p^2} \times C_p \times C_p$ is a non-elementary abelian p-group with $p^2 + p + 1$ subgroups of order p.

Chapter 15

1. (a) The group $S(4)$ has $A(4)$ as a normal subgroup of index two. It follows that a composition series is

$$S(4) \geq A(4) \geq V \geq \{1, (12)(34)\} \geq \{1\}.$$

This is not a chief series because none of the three subgroups of V with two elements is normal in $S(4)$. Since V is itself normal in $S(4)$, a chief series is

$$S(4) \geq A(4) \geq V \geq \{1\}.$$

(b) Let G be the quaternion group of order 8. Since $\langle a \rangle$ has index 2, it is normal. The subgroup $\langle a^2 \rangle$ has order 2 and, since $a^2 = b^2$, this element commutes with every element of G. It follows that $\langle a^2 \rangle$ is a normal subgroup and that the series

$$G \geq \langle a \rangle \geq \langle a^2 \rangle \geq \{1\}$$

is both a chief series and a composition series.

(c) The dihedral group $G = D(6)$ has 12 elements. Let a be an element of order 6, and b be an element of order 2 with $bab^{-1} = a^{-1}$. Then $\langle a \rangle$ is a normal subgroup of G, being of index two. The subgroup $\langle a^2 \rangle$ is a subgroup of $\langle a \rangle$ of order 3, and is a normal subgroup of $\langle a \rangle$, since $\langle a \rangle$ is abelian. Thus

$$G \geq \langle a \rangle \geq \langle a^2 \rangle \geq \{1\}$$

is a composition series for G. This is also a chief series because $\langle a^2 \rangle$ is actually a normal subgroup of G. To see this, notice that

$$ba^2b^{-1} = (bab^{-1})^2 = (a^{-1})^2 = a^{-2} \in \langle a^2 \rangle.$$

Since b and a are both in the normaliser $N_G\langle a^2 \rangle$, G is also in $N_G\langle a^2 \rangle$ and so $\langle a^2 \rangle$ is a normal subgroup of G.

2. A composition series for $A(4) \times A(4)$ is

$$\begin{aligned} A(4) \times A(4) \quad &\geq \quad A(4) \times V \geq A(4) \times \{1, (1\ 2)(3\ 4)\} \\ &\geq \quad A(4) \times \{1\} \geq V \times \{1\} \\ &\geq \quad \{1, (1\ 2)(3\ 4)\} \times \{1\} \geq \{1\} \times \{1\}, \end{aligned}$$

because each subgroup is normal in the one above, by Question 2 of Exercises 11, and the indices are all primes. A chief series is

$$A(4) \times A(4) \geq A(4) \times V \geq A(4) \times \{1\} \geq V \times \{1\} \geq \{1\} \times \{1\},$$

since none of the three subgroups of V of order 2 is a normal subgroup of $A(4)$.

3. Examples are provided by

(a) the infinite cyclic group;
(b) the group $A(4)$; and
(c) the cyclic group of order 6.

Chapter 16

1. The group C_6 has a chief series in which the factors are C_2 and C_3, as does the symmetric group $S(3)$.

2. The normal subgroups of $S(3)$ are $S(3)$, $A(3)$ and $\{1\}$, and since these are each the unique subgroup of their orders, they are all characteristic subgroups.

In $D(4)$, the subgroup $\{1, a^2\}$ is the centre of $D(4)$. Since any automorphism must map a central element onto a central element, this subgroup is characteristic. None of the other three subgroups with two elements (that is, (1,1) together with one of the three elements $\{1, b\}, \{1, ba\}$ or $\{1, ba^2\}$) is characteristic since none is normal in G. Among the subgroups with four elements, $\langle a \rangle$ is characteristic since an automorphism must map any element of order 4 onto an element of order 4. However the other two subgroups of order 4,

$$\{1, b, a^2, ba^2\} \text{ and } \{1, ba, a^2, ba^3\},$$

are not since the map determined by $a \mapsto a$ and $b \mapsto ba$ determines an automorphism of G.

Finally in Q_8, there is only one subgroup of order 2, so it is characteristic, but there are three normal subgroups of order 4 which are permuted amongst themselves by an automorphism of Q.

3. An example is 60. The alternating group $A(5)$ has a composition series $A(5) \geq \{1\}$ with only one composition factor. The cyclic group of order 60 has a composition series

$$C_{60} \geq C_{30} \geq C_{15} \geq C_5 \geq \{1\}$$

with 4 composition factors.

4. Let ϑ be an automorphism of G. Since H is a characteristic subgroup of G, H is invariant under ϑ, so that $\vartheta(h) \in H$ for all $h \in H$. It follows that ϑ induces an automorphism of H by restricition. Since K is a characteristic subgroup of H, K is invariant under this restricted map, so that $\vartheta(K) = K$, as required.

Chapter 17

1. For $n \leq 4, S(n)$ is soluble, since $S(1)$ has order $1, S(2)$ is cyclic of order 2, while $S(3)$ and $S(4)$ are considered in the text. For $n \geq 5$, the normal subgroup $A(n)$ is simple by Theorem 16.16, so $S(n) \geq A(n) \geq \{1\}$ is a composition series for G. It then follows by the Jordan–Hölder Theorem that $S(n)$ cannot have a composition series with abelian factors.

2. (a) It may be checked that

$$\begin{aligned} (\vartheta(G))' &= \langle [\vartheta(x), \vartheta(y)] : x, y \in G \rangle \\ &= \langle \vartheta([x, y]) : x, y \in G \rangle = \vartheta(G'). \end{aligned}$$

(b) If H is soluble and ϑ is injective then $\ker \vartheta = \{1\}$, and so by the Homomorphism Theorem $G/\ker \vartheta = G \cong \operatorname{im} \vartheta$. Since $\operatorname{im} \vartheta$ is a subgroup of the soluble group H, we deduce that G is soluble.

(c) If G is soluble and ϑ is surjective then im $\vartheta = H$ and so the homomorphism theorem implies that $G/\ker\vartheta$ is isomorphic to H. Thus H is isomorphic to a quotient group of a soluble group and so is soluble.

3. We first show that G is a group. Let $a, b, c, d \in \mathbf{R}$ with $a \neq 0$ and $c \neq 0$. Then

$$\vartheta_{a,b}\vartheta_{c,d}(x) = \vartheta_{a,b}(cx + d) = a(cx + d) + b = \vartheta_{ac,ad+b}(x).$$

Since $ac \neq 0, G$ is closed. Associativity follows from associativity of maps. The identity element is $\vartheta_{1,0}$. The inverse of $\vartheta_{a,b}$ is $\vartheta_{1/a,-b/a}$. It follows that the commutator $[\vartheta_{a,b}, \vartheta_{c,d}]$ is of the form $\vartheta_{1,e}$ for some e. It is now easily checked that the subset $\{\vartheta_{1,e} : e \in \mathbf{R}\}$ is an abelian subgroup of G, so the commutator subgroup of G is abelian. Thus $G^{(2)} = \{1\}$ and hence G is soluble by Proposition 17.15.

4. (a) The following calculation gives the result:

$$\begin{aligned}
[ab, c] &= (ab)c(ab)^{-1}c^{-1} \\
&= abcb^{-1}a^{-1}c^{-1} \\
&= a(bcb^{-1}c^{-1})ca^{-1}c^{-1} \\
&= a[b, c]a^{-1}aca^{-1}c^{-1} = [b, c]^a[a, c].
\end{aligned}$$

(b) Similarly

$$\begin{aligned}
[a, bc] &= abca^{-1}c^{-1}b^{-1} \\
&= aba^{-1}b^{-1}baca^{-1}c^{-1}b^{-1} \\
&= [a, b][a, c]^b.
\end{aligned}$$

5. The additional relations that $t_it_j = t_jt_i$ for all i, j imply that $t_it_{i+1} = 1$. Thus, $t_i = t_{i+1}$, so that the commutator quotient group of $S(n)$ is cyclic of order 2.

6. Since

$$\begin{pmatrix} 1 & a & b & c \\ 0 & 1 & 0 & d \\ 0 & 0 & 1 & e \\ 0 & 0 & 0 & 1 \end{pmatrix}\begin{pmatrix} 1 & \alpha & \beta & \gamma \\ 0 & 1 & 0 & \delta \\ 0 & 0 & 1 & \epsilon \\ 0 & 0 & 0 & 1 \end{pmatrix}$$

$$= \begin{pmatrix} 1 & a+\alpha & b+\beta & c+b\epsilon+a\delta+\gamma \\ 0 & 1 & 0 & d+\delta \\ 0 & 0 & 1 & e+\epsilon \\ 0 & 0 & 0 & 1 \end{pmatrix}$$

we see that G is closed under multiplication. Also

$$\begin{pmatrix} 1 & a & b & c \\ 0 & 1 & 0 & d \\ 0 & 0 & 1 & e \\ 0 & 0 & 0 & 1 \end{pmatrix}^{-1} = \begin{pmatrix} 1 & -a & -b & -c+be+ad \\ 0 & 1 & 0 & -d \\ 0 & 0 & 1 & -e \\ 0 & 0 & 0 & 1 \end{pmatrix}.$$

It is now clear that G is a group, and it is easily checked that A is an abelian normal subgroup of G. The quotient group G/A is abelian, and so G' is contained in A, showing that G is soluble.

Chapter 18

1. The symmetric group $S(3)$ is an example of a non-nilpotent group with a nilpotent normal subgroup, $A(3)$, with nilpotent quotient group.

2. It can be easily checked that for any A in G,

$$A = \begin{pmatrix} 1 & a & b \\ 0 & 1 & c \\ 0 & 0 & 1 \end{pmatrix}, \ A^2 = \begin{pmatrix} 1 & 2a & 2b+ac \\ 0 & 1 & 2c \\ 0 & 0 & 1 \end{pmatrix},$$

so that

$$A^3 = \begin{pmatrix} 1 & 3a & 3b+3ac \\ 0 & 1 & 3c \\ 0 & 0 & 1 \end{pmatrix}.$$

Since $3x = 0$, we see that $A^3 = I$, so that every non-identity element has order 3. Writing down the condition for two matrices in G to commute, it can be easily seen that the centre $Z(G)$ consists of the three matrices

$$\begin{pmatrix} 1 & 0 & 0 \\ 0 & 1 & 0 \\ 0 & 0 & 1 \end{pmatrix}, \begin{pmatrix} 1 & 0 & 1 \\ 0 & 1 & 0 \\ 0 & 0 & 1 \end{pmatrix} \text{ and } \begin{pmatrix} 1 & 0 & 2 \\ 0 & 1 & 0 \\ 0 & 0 & 1 \end{pmatrix}.$$

It may be checked that the nine matrices in G with $c = 0$ form a subgroup, N. Since this subgroup is a maximal subgroup of the p-group G, it is normal and so

$$G \geq N \geq Z(G) \geq \{1\}$$

is a normal series for G with quotients cyclic of order 3, and so is a chief series for G.

3. Since $\langle x \rangle$ has nine elements, and so has index 3, $\langle x \rangle$ is a normal subgroup of P by Corollary 18.7. Also

$$yx^3y^{-1} = (yxy^{-1})^3 = (x^4)^3 = x^{12} = x^3,$$

so y centralises x^3. Since x centralises x^3, P centralizes x^3 and so x^3 is in the centre of P. It follows that

$$P \geq \langle x \rangle \geq \langle x^3 \rangle \geq \{1\}$$

is a chief series for P.

4. The groups C_{27}, $C_9 \times C_3$ and $C_3 \times C_3 \times C_3$ are the abelian groups of order 27, and are non-isomorphic by the classification theorem for finite abelian groups. The other two are non-abelian, so cannot be isomorphic to any abelian group. They are not isomorphic to each other because the group G in Exercise 2 has no element of order 9. In G and $C_3 \times C_3 \times C_3$ there are 26 non-identity elements of order 3.

5. The integers 61, 67 and 71 are primes, 62, 65 and 69 are a product of two primes, 63 and 68 are of the form p^2q, and 64 is a prime power. This leaves 66 and 70. A group with 66 elements must have a normal Sylow 11-subgroup and a group of order 70 must have a normal Sylow 7-subgroup, so all groups with either of these orders are soluble.

Chapter 19

1. Recall that A rwr B has order $|A|^{|B|}|B|$, so $(G \text{ rwr } H)$ rwr K has order

$$(|G|^{|H|}|H|)^{|K|}|K| = |G|^{|H||K|}|H|^{|K|}|K|.$$

On the other hand, G rwr $(H$ rwr $K)$ has order

$$|G|^{|H|^{|K|}|K|}|H|^{|K|}|K|.$$

Since these integers are different, the groups cannot be isomorphic.

2. The number of conjugates of $\rho = (1\ 2)(3\ 4)(5\ 6)$ in $S(6)$ is $6!/2^3 3! = 15$, so the centraliser of ρ has order $2^3 3! = 48$. The elements in the group

$$N = \langle (1\ 2) \rangle \times \langle (3\ 4) \rangle \times \langle (5\ 6) \rangle$$

all centralise ρ, as do the permutations $\sigma = (1\,3)(2\,4)$ and $\tau = (1\,3\,5)(2\,4\,6)$. However, σ and τ generate a subgroup isomorphic to $S(3)$, and so the group $\langle N, \sigma, \tau \rangle$ has 48 elements, all of which centralise ρ. It may be easily checked that $\langle N, \sigma, \tau \rangle$ is isomorphic to C_2 pwr $S(3)$.

3. Let P be a Sylow p-subgroup of G of order p^r, say. Suppose now that Q is a Sylow q-subgroup of G, so that $|Q| = q^s$, where $|G| = p^r q^s$. It is clear that $P \cap Q = \{1\}$, since this is a subgroup of both P and Q. Since P is a

normal subgroup of G and $P \cap Q = \{1\}$, $|PQ| = |P||Q| = |G|$. Thus G is a semidirect product of P and Q.

4. Suppose that G is the semidirect product of N and H, so that every element of G is of the form nh, for $n \in N$ and $h \in H$. Since N has two elements, 1 and z, say, for any h in H, the conjugate hzh^{-1} is in N and cannot equal 1. It follows that z (and therefore) N is in the centre of G. Thus, H is a normal subgroup of G: for any h_0 in H, and $g = nh$ in G,

$$gh_0g^{-1} = nhh_0(nh)^{-1} = nhh_0h^{-1}n = hh_0h^{-1} \in H$$

since N is in the centre of G. Thus N and H are normal subgroups of G with $N \cap H = \{1\}$ and $G = NH$, so G is an internal direct product of N and H.

5. Any automorphism of $G = C_3 = \langle x \rangle$ is determined by its value on x. The only possibiltites are the trivial map and the map $x \mapsto x^2 = x^{-1}$. Since neither of these maps has order 2, it follows that the only homomorphism $\mathrm{Aut}(G) \to C_3$ is the identity map. It now follows by Proposition 19.4 that the extended group is abelian of order nine. Since the extended group is a semidirect product, it is isomorphic to $C_3 \times C_3$.

Chapter 20

1. Possible sections are of the form {1, odd permutation}, so there are three sections: $\{1, (12)\}, \{1, (13)\}$ and $\{1, (23)\}$. Since each of these sections is a subgroup, the corresponding sectional factor sets all take the value 1 everywhere.

2. The required sectional factor set is

	1	(1 2 3)	(1 2 4)
1	1	1	1
(1 2 3)	1	(1 4)(2 3)	(1 3)(2 4)
(1 2 4)	1	(1 4)(2 3)	(1 3)(2 4)

An example of a section which is a subgroup is $\{1, (1\ 2\ 3), (1\ 3\ 2)\}$.

3. The cyclic group of order 2 only has one automorphism, so the extension in this case is also a direct product, $C_2 \times C_3 \cong C_6$.

Chapter 21

1. Let G consist of the eight elements $t(1)$, $t(a)$, $t(b)$, $t(ab)$, $zt(1)$, $zt(a)$, $zt(b)$, $zt(ab)$. Writing x for $t(a)$, and y for $t(b)$, we see that

$$x^2 = f(a,a)t(1) = 1, \; y^2 = f(b,b)t(1) = z,$$

and

$$xy = zt(ab) = zyx = y^{-1}x,$$

so G is isomorphic to $D(4)$.

2. Let G be a central extension of $C_3 = \langle z \rangle$ by $S(3)$. Then G has a subgroup K of index 2 which is a central extension of C_3 by $A(3)$. It follows that K is abelian, and so K is either $C_3 \times C_3$ or C_9. Let y be an element of order 2 in G, so that y is a generator for a Sylow 2-subgroup of G. Suppose that K is cyclic generated by x, so that $x^3 = z$. Since $G/\langle z \rangle$ is $S(3)$, $yxy^{-1} = x^{-1+3k}$. But then y would conjugate $x^3 = z$ to $x^{-3+9k} = z^{-1}$, contrary to the fact that z is central. This case cannot therefore arise. We therefore suppose that K is non-cyclic, generated by z and x, say. Renaming the elements of K if need be, we therefore see that G has presentation

$$\langle x, y, z : x^3 = y^2 = z^3 = 1, \; xz = zx, \; yz = yz, \; yxy^{-1} = x^{-1} \rangle,$$

so that G is $S(3) \times C_3$.

3. The only non-trivial automorphism of the infinite cyclic group G generated by x is determined by $x \mapsto x^{-1}$. It follows that the extensions of G by C_2 are isomorphic to $G, G \times C_2$ or to the infinite dihedral group

$$\langle x, y : y^2 = 1, yxy^{-1} = x^{-1} \rangle.$$

Chapter 22

1. Suppose that G has order 18. Let P be a Sylow 3-subgroup of order 9. Since P has index 2, it is a normal subgroup of G. We know that P is either cyclic or $C_3 \times C_3$. Let $Q = \langle y \rangle$ be a Sylow 2-subgroup. If P is cyclic generated by x, then $yxy^{-1} = x^i$, for some i. Since $y^2 = 1$, $i^2 \equiv 1 \bmod 9$. It may be seen by inspection that either $i \equiv 1 \bmod 9$ (in which case G is abelian), or $i \equiv -1 \bmod 9$ (in which conjugation by y has no non-identity fixed points, and so G is isomorphic to $D(9)$).

We may therefore suppose that $P = \langle a, b \rangle = C_3 \times C_3$. By Proposition 22.4 the automorphism group of P is isomorphic to the group $GL(2,3)$. The elements of order 2 in $\mathrm{Aut}(P)$ are $-I$ (the only matrix with characteristic polynomial $(\lambda - 1)^2$) together with 12 matrices conjugate to $\begin{pmatrix} 1 & 0 \\ 0 & -1 \end{pmatrix}$ (whose eigenvalues are 1 and -1). Since y conjugates P according to one of these two types of automorphism, we see that G is isomorphic either to $D(3) \times C_3$, or to the group

$$\langle x, y, z : x^3 = y^3 = z^2 = 1,\ xy = yx,\ zxz^{-1} = x^{-1},\ zyz^{-1} = y^{-1} \rangle.$$

2. It is clear from Sylow theory that G has a normal Sylow 5-subgroup, P, say. Let $P = \langle x \rangle$. The automorphism group of P is cyclic of order 4, and no non-identity automorphism has a fixed point. Thus if a Sylow 2-subgroup of G is cyclic, generated by y, and if G is non-abelian, we have the possible presentations:

$$\langle x, y : x^5 = y^4 = 1,\ yxy^{-1} = x^2 \rangle;$$

$$\langle x, y : x^5 = y^4 = 1,\ yxy^{-1} = x^3 \rangle;$$

$$\langle x, y : x^5 = y^4 = 1,\ yxy^{-1} = x^{-1} \rangle.$$

The first two groups are isomorphic (replacing y by y^{-1}) and are each isomorphic to the holomorph of C_5. Putting $u = xy^2$ and $v = y$, we see that the third group is the quaternionic group, Q_{20}, generated by u and v. If a Sylow 2-subgroup is non-cyclic, generated by a and b, say, with a inverting x and b centralising x, it may be checked that G is the dihedral group $D(10)$.

3. Let $P = \langle x \rangle$ be the normal Sylow 7-subgroup of G, so that $\mathrm{Aut}(P)$ has order 6. It follows that $C_G(P)$ has index at most 2, and so, if G is non-abelian, G has a cyclic subgroup of order 14. If y is an element outside this subgroup, y inverts a generator and so G is either $D(14)$ or Q_{28}.

4. Suppose that G has order 70. It is clear that G has a unique Sylow 5-subgroup and a unique Sylow 7-subgroup, and therefore has a normal subgroup of order 35 and index 2. A group of order 35 is cyclic by Exercise 12.1. If x is a generator for this subgroup of order 35, and y is a generator for a Sylow 2-subgroup, $yxy^{-1} = x^i$, where $i^2 \equiv 1 \bmod 35$. It can be checked by inspection that the only solutions of this congruence are $i \equiv 1, 6, 29, 34$ modulo 35. These four solutions lead to the groups C_{70}, $C_5 \times D(7)$, $C_7 \times D(5)$ and $D(35)$.

5. The automorphism group of $C_2 \times C_2$ is $GL(2, 2)$ of order 6. The elements of order 2 in this group are

$$\begin{pmatrix} 1 & 1 \\ 0 & 1 \end{pmatrix}, \begin{pmatrix} 1 & 0 \\ 1 & 1 \end{pmatrix} \text{ and } \begin{pmatrix} 0 & 1 \\ 1 & 0 \end{pmatrix}.$$

Choosing generators a and b for the normal subgroup $C_2 \times C_2$, we see that either G is abelian (in which case G is either $C_4 \times C_2$ or $C_2 \times C_2 \times C_2$) or G has one of the presentations

$$\langle a, b, c : a^2 = b^2 = c^2 = 1,\ ab = ba,\ cac^{-1} = a,\ cbc^{-1} = ab\rangle \text{ or}$$

$$\langle a, b, c : a^2 = b^2 = 1,\ c^2 = a,\ ab = ba,\ cac^{-1} = a,\ cbc^{-1} = ab\rangle.$$

Both these groups are isomorphic to the dihedral group $D(4)$.

6. The dihedral group $D(6)$, generated by a of order 6 and b of order 2, has a normal subgroup generated by a^2 and b isomorphic to $S(3)$, and a normal subgroup, generated by a^3, isomorphic to C_2. These intersect in $\{1\}$, so it follows that $D(6)$ is the internal direct product $S(3) \times C_2$.

Bibliography

Anderson, I. (1989). *A first course in combinatorial mathematics.* Oxford Applied Mathematics and Computing Science Series. Clarendon Press, Oxford.

Artin, E. (1957). *Geometric algebra.* Wiley, New York.

Aschbacher, M. (1994). *Sporadic groups.* Cambridge Tracts in Mathematics 104. Cambridge University Press.

Cartan, É. (1952). Sur la structure des groupes de transformations finis et continus. *Oeuvres completes* Vol.1, pp. 137–285.

Carter, R.W. (1985). *Finite groups of Lie type.* Wiley, New York.

Cayley, A. (1854). On the theory of groups, as depending on the symbolic equation $\theta^n = 1$. *Philos. Magazine Royal Society London*, **7**, 40–7.

Cayley, A. (1854). The theory of groups. *Amer. J. Math.*, **1**, 50–2.

Chevalley, C. (1955). Sur certaines groupes simples. *Tohoku Math. J.*, **7**, 14–66.

Conway, J.H., Curtis, R.T., Norton, S.P., Parker, R.A., Wilson, R.A. (1985). ATLAS *of finite groups.* Clarendon Press, Oxford.

Dickson, L.E. (1900). *Linear groups with an exposition of the Galois field theory.* Teubner, Leipzig.

Dieudonné, J. (1948). *Sur les groupes classiques.* Hermann, Paris.

Gorenstein, D. (1982). *Finite simple groups: an introduction to their classification.* Plenum Press, New York.

Hill, R. (1986). *A first course in coding theory.* Oxford Applied Mathematics and Computing Science Series. Clarendon Press, Oxford.

Hölder, O. (1889). Zurückführung einer beliebigen algebraischen Gleichung auf eine Kette von Gleichungen. *Math. Ann.*, **34**, 26–56.

Humphreys, J.F. and Prest, M.Y. (1989) *Numbers, groups and codes.* Cambridge University Press.

Jordan, C. (1869). Commentaire sur Galois. *Math. Ann.*, **1**, 141–60.

Jordan, C. (1870). *Traité des substitutions et des équations algébriques.* Gauthier-Villars, Paris.

Kaloujnine, L. (1948). La structure des p-groupes de Sylow des groupes symétriques finis. *Ann. Sci. École Norm. Super.*, **65**, 239–76.

Kronecker, L. (1870). Auseinandersetzung einiger Eigenschaften der Klassenzahl idealer komplexer Zahlen. *Monatsberichte der Berliner Akademie*, 881–9.

Mathieu, E. (1861). Mémoire sur l'étude des fonctions de plusiers quantités. *J. Math. Pures et Appl.*, **6**, 241–323.

Mathieu, E. (1873). Sur la fonction cinq fois transitives de 24 quantités. *J. Math. Pures et Appl.*, **18**, 25–46.

Miller, G.A. (1900). Sur plusieurs groupes simples. *Bull. Soc. Math. de France*, **28**, 266–7.

Rose, J.S. (1978). *A course on group theory.* Cambridge University Press.

Schur, I. (1904). Über die Darstellung der endlichen Gruppen durch gebrochene lineare Substitutionen. *J. für Math.*, **127**, 20–50.

Scott, W.R. (1964.) *Group theory.* Prentice-Hall, New Jersey.

Schreier, O. (1926a). Über die Erweiterung von Gruppen, I. *Monatshefte für Mathematik und Physik*, **34**, 165–80.

Schreier, O. (1926b). Über die Erweiterung von Gruppen, II. *Abhandlungen Math. Seminar Hamburg*, **4**, 321–46.

Schreier, O. (1928). Über den Jordan–Höldersche Satz. *Abhandlungen Math. Seminar Hamburg*, **6**, 300–2.

Steinberg, R. (1959). Variations on a theme of Chevalley. *Pacific J. Math.*, **9**, 875–91.

Suzuki, M. (1982) *Group theory I.* Springer-Verlag, Berlin.

Sylow, M.L. (1872). Théorèmes sur les groupes de substitutions. *Math. Ann.*, **5**, 584–94

van der Waerden, B.L. (1985). *A history of algebra.* Springer-Verlag, Berlin.

Wielandt, H. (1954). Zum Satz von Sylow. *Math. Z.*, **60**, 407–9.

Zassenhaus, H. (1934). Zum Satz von Jordan–Hölder–Schreier. *Abhandlungen Math. Seminar Hamburg*, **10**, 106–8.

Index